The Resilience Assessment Framework

Assessing Commercial Contributions to U.S. Space Force Mission Resilience

OSONDE A. OSOBA, GEORGE NACOUZI, JEFF HAGEN, JONATHAN TRAN, LI ANG ZHANG, MARISSA HERRON, CHRISTOPHER LYNCH, MEL EISMAN, CHARLES BARTON

Prepared for the Department of the Air Force
Approved for public release; distribution unlimited

PROJECT AIR FORCE

For more information on this publication, visit **www.rand.org/t/RRA980-1**.

Published by the RAND Corporation, Santa Monica, Calif.

Library of Congress Cataloging-in-Publication Data is available for this publication.

ISBN: 978-1-9774-1074-0

Cover: Alex/Adobe Stock and photo courtesy of United Launch Alliance.

About This Report

The Department of the Air Force (DAF) and the U.S. Space Force (USSF) are exploring opportunities to use commercial space services to augment government capabilities. Decisions to purchase commercial services need to be evaluated along many dimensions, one of which is mission performance, particularly resilience. To help the USSF conduct such assessments, RAND Project AIR FORCE (PAF) developed a resilience assessment framework that allows the user to compare government architectures, commercial architectures, and combined government and commercial architecture along relevant measures of performance and to understand how these architectures respond under various levels of degradation. The framework also incorporates an assessment of the trustworthiness of both the commercial provider and the information provided. This report describes the overall framework and discusses technical details of the RAND-developed methodologies contained within the framework. Policy-level findings and recommendations using this framework can be found in a companion report.[1] The intended audience for this report includes USSF personnel developing and analyzing space architectures, as well as DAF space personnel interested in space-based surveillance.

The research reported here was commissioned by the USSF and conducted within the Force Modernization and Employment Program of RAND Project AIR FORCE as part of a fiscal year 2021 project, "Enhancing Space Mission Survivability and Resilience by Leveraging Commercial Space Capabilities."

RAND Project AIR FORCE

RAND Project AIR FORCE (PAF), a division of the RAND Corporation, is the Department of the Air Force's (DAF's) federally funded research and development center for studies and analyses. PAF provides the DAF with independent analyses of policy alternatives affecting the development, employment, combat readiness, and support of current and future air, space, and cyber forces. Research is conducted in four programs: Strategy and Doctrine; Force Modernization and Employment; Resource Management; and Workforce, Development, and Health. The research reported here was prepared under contract FA7014-16-D-1000.

Additional information about PAF is available on our website:

www.rand.org/paf/

---

[1] George Nacouzi, Osonde A. Osoba, Jeff Hagen, Jonathan Tran, Christopher Lynch, Mel Eisman, Li Ang Zhang, Charlie Barton, Marissa Herron, and Yool Kim, *Commercial Space Services: An Opportunity for U.S. Space Force to Increase its Mission Resilience*, RAND Corporation, forthcoming, Not available to the general public.

This report documents work originally shared with the DAF on September 29, 2021. The draft report, issued September 30, 2021, was reviewed by formal peer reviewers and DAF subject-matter experts.

Acknowledgments

This research was made possible through the support of our USSF sponsor, Brig Gen Kevin Whale. We also want to give special thanks to Lt Col Erik Bowman who provided valuable and insightful information on the use of commercial space by USSF.

We also want to thank the many government subject-matter experts representing the Space Warfighting Analysis Center, Air Force Research Laboratory, and the National Reconnaissance Office commercial space office who provided valuable and insightful feedback and suggestions.

We thank Jim Chow at RAND for his guidance and support for this research. We are also grateful to Barbara Bicksler for her crucial assistance in the preparation of this report.

Summary

Over the past few years, commercial space services have significantly increased in capability and capacity in many missions of interest to the U.S. Space Force (USSF). As the USSF considers incorporating such commercial space services into its missions, it needs a principled and flexible assessment framework for evaluating how commercial contributions affect the performance and resilience of various USSF missions.

This report describes the resilience assessment framework that RAND Project AIR FORCE (PAF) developed to assess the effects of commercial contributions on the resilience of USSF missions and shows how the authors applied the framework to evaluate commercial space contributions in two sample USSF missions.

Issue

The U.S. government has stated its intent to leverage new commercial space capability, and the Department of the Air Force (DAF) has been assessing the use of some of these capabilities. However, there is a need to develop and apply a principled quantitative framework for evaluating the effects of proposed commercial contributions on the resilience of USSF missions.

Approach

We developed an assessment framework that combines physics-based modeling and statistical systems analyses to help assess the impact of select commercial services on USSF space-based mission performance. The framework considers the additional mission performance and mission resilience that a proposed commercial service could provide. It also includes a multilayered approach to grade information contributions from commercial services for trustworthiness.

The framework can be tailored to diverse missions by specifying the *relevant* mission assets or infrastructure, the commercial services available, and the mission performance measures. We used the framework to assess the commercial contributions for two specific example missions: tactical intelligence, surveillance, and reconnaissance (T-ISR) and data transmit and receive network (DTRN).

Findings

We applied the resilience assessment framework to the T-ISR and DTRN case studies. The results show that the framework is highly flexible and useful for providing insight into how the performance and resilience of whole mission systems respond to adaptations of their infrastructure, including infrastructure augmentation with commercial services, as we explored in this study. As an example, Figure S.1 shows a sample result from our assessment of the value of commercial services on the resilience of the DTRN mission.

The results depict such trends as the relative rates of performance decline for USSF-only versus the USSF + Commercial setups for both case-study missions. The results also provide estimates of how much degradation mission systems can withstand before their performance falls below prespecified limits. This allows analysts to compare degradation thresholds across USSF-only versus the USSF + Commercial mission systems. We also note that to effectively leverage commercial capabilities, the USSF will need to consider the entire concept of operations (CONOPS) associated with using these capabilities.[2]

Figure S.1. Comparison of Average Satellite Revisit Time, by Mission System

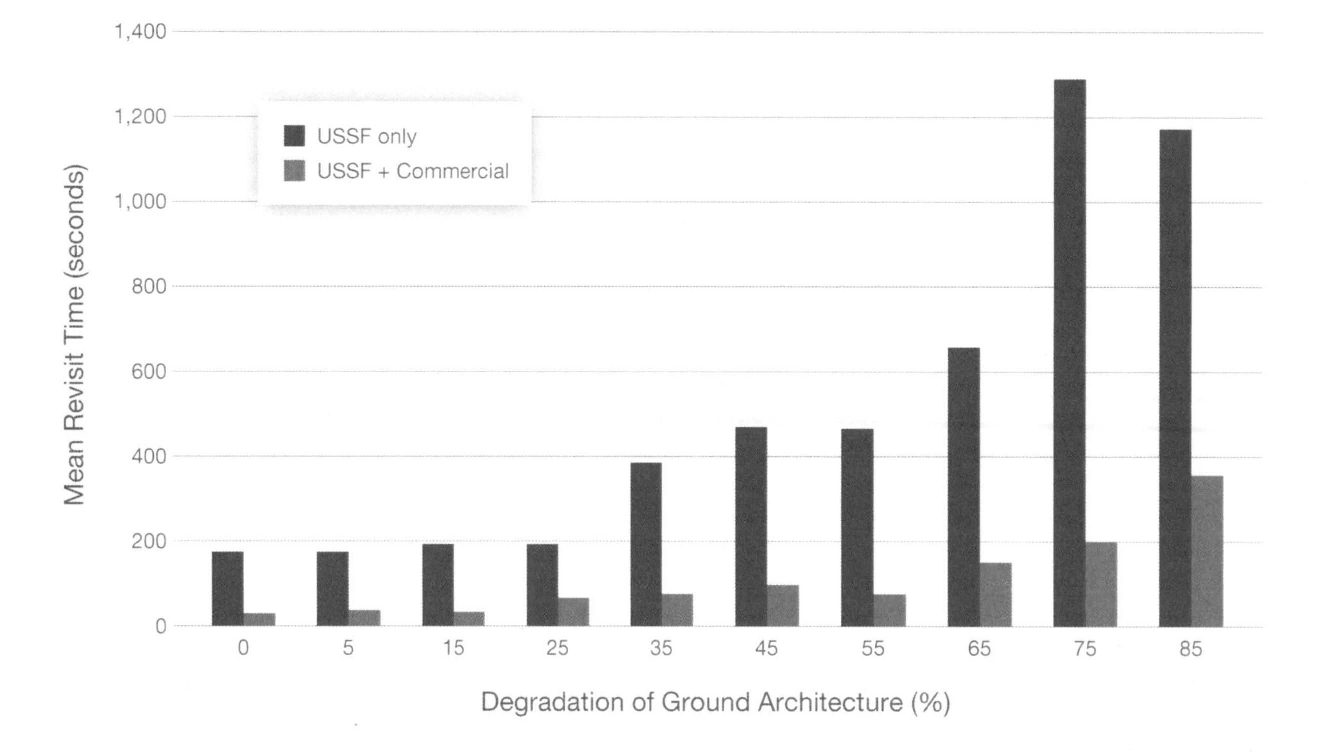

Recommendations

We identified the following recommendations for the USSF to consider:

- *Extend the framework to simulate responses to adversarial targeted degradation*: Extending the shock simulation in the resilience assessment framework will enable the USSF to evaluate mission systems' response to more targeted *adversarial* system

[2] We did not investigate the CONOPS modifications needed to leverage commercial capabilities. That should be the subject of future research.

degradation. Adversarial targeted degradation contrasts with purely random degradation that our work assumed (i.e., we degraded the system by randomly excluding some satellites).

- *Further develop and implement the trust assessment subframework*: We identified a few viable approaches for "fingerprinting" commercially sourced information artifacts for security and chain-of-custody tracking. And we identified a promising approach for fusing together trust signals into a summary metric of trust. However, there is further work to do to implement the approach from end to end for a specific mission.
- *Use improved representations of USSF mission systems in resilience analyses*: We make simplifying approximations of the missions for the purposes of this study; however, a higher fidelity representation should be used in future work.
- *Operationalize an appropriately tailored resilience assessment framework for USSF missions*: The resilience and trust requirements of USSF missions demand caution when considering integrating commercial services. The USSF (and the DAF more generally) needs a *principled*, *quantitative* framework for assessing the contributions of imminent or proposed commercial augmentations to space-based defense missions. Our resilience assessment framework serves this purpose and is designed to be modular and flexible.
- *Explore applications of the framework to other missions*: The motivating idea behind our resilience assessment framework is not limited in relevance to just missions in space. The application of the proposed framework to other missions should be considered.
- *Assess what modifications to the overall USSF CONOPS are required to effectively leverage commercial capabilities*: This should involve assessing the impact of integrating commercial capabilities with USSF mission systems on doctrine, organization, training, materiel, leadership, personnel, facilities, and policy.

Contents

Figures and Tables

Figures

Tables

Chapter 1. Overview of the Resilience Analysis Framework

As commercial space services increase in capability and capacity, more opportunities are available for the U.S. government to use commercial services to augment organic capabilities. The Department of the Air Force is already experimenting with the use of some of these capabilities. But a key question is how much and which of the new evolving space services should the government purchase? The answer to this question is multifaceted, but one important dimension is an understanding of the impact of commercial services on mission performance, particularly resilience. Because of the importance of resilience to its missions, the U.S. Space Force (USSF) asked RAND Project AIR FORCE (PAF) to help explore this topic.

Building on past RAND Corporation work,[3] we developed a resilience assessment framework to evaluate the performance impact of leveraging new commercial space capabilities, which, along with a deeper dive into the methodologies underlying the framework, is described in this report. We begin in this chapter with an overview of the framework. Later chapters give more details on various elements of the framework and provide illustrative examples of how the framework applies to sample USSF missions.

Framework for Conceptualizing and Assessing Resilience

We can situate our resilience approach in the context of prior thinking about resilience analysis. Prior works identify useful baseline concepts for evaluating resilience.[4] One of the recurring themes in this body of work is the conceptualization of resilience in terms of *a system's sensitivity to changes* either in the operating environment (e.g., adversarial threats or just random events) or in the system's structure (e.g., augmentation or degradation). There is not full consensus on the exact system metric to measure resilience in the face of changes most likely because such metrics are application and system dependent. Thus, we anchor our analysis to evaluating a system's sensitivity to changes rather than to any one prescribed system measure.

We use two relevant frames for quantifying resilience under different system measures: (1) a *performance threshold account of resilience* and (2) an *elasticity account of resilience*.

---

[3] Yool Kim, George Nacouzi, Mary Lee, Brian Dolan, Krista Romita Grocholski, Emmi Yonekura, Moon Kim, Thomas Light, and Raza Khan, *Leveraging Commercial Space Capabilities to Enhance the Space Architecture of the U.S. Department of Defense*, RAND Corporation, 2022, Not available to the general public.

[4] Paul Dreyer, Krista S. Langeland, David Manheim, Gary McLeod, and George Nacouzi, *RAPAPORT (Resilience Assessment Process and Portfolio Option Reporting Tool): Background and Method*, RAND Corporation, RR-1169-AF, 2016; Graeme B. Shaw, David W. Miller, and Daniel E. Hastings, "Development of the Quantitative Generalized Information Network Analysis Methodology for Satellite Systems," *Journal of Spacecraft and Rockets*, Vol. 38, No. 2, March–April 2001; Ron Burch, *Resilient Space Systems Design: An Introduction*, CRC Press, 2019.

Performance threshold account of resilience asks: "How much change or degradation (Δx) can a system handle before it no longer performs, $P(\Delta x)$, at or above a required threshold level of performance (i.e., $P_{threshold}$)?" This performance threshold account is useful especially when mission context has prespecified performance standards against which to measure and is expressed as follows:

$$\text{argmin}_{\Delta x} P(\Delta x) \leq P_{threshold}.$$

The elasticity account of resilience asks: "What is the change in system performance (ΔP) typically associated with a change in the system's input or configuration (Δx)?" This account allows the analyst to specify an elasticity-like quantity that summarizes or estimates (roughly) a system's resilience and is expressed as follows:[5]

$$R \approx \frac{\Delta x}{\Delta P}.$$

This quantity is useful for comparing systems without the use of a prespecified performance threshold. Both conceptions of resilience will feature in our analyses.

From a more practical perspective, the U.S. Department of Defense (DoD) describes resilience as the

> ability of an architecture to support the functions necessary for mission success with higher probability; shorter periods of reduced capability; and across a wider range of scenarios, conditions, and threats, in spite of hostile action or adverse conditions.[6]

However, a quantitative operationalization of resilience assessments can take many forms (as the many articles, papers, and even a dedicated book on the topic can attest).[7] Joint Publication (JP) 3-14 includes resilience as a subset of mission assurance; reconstitution and defensive operations are the other two subsets.[8] In this report, we focus on only resilience at the request of the sponsor and to keep within scope. JP 3-14 (2020) also provides six different methods that contribute to resilience: disaggregation, distribution, diversification, protection, proliferation, and deception.[9] Although the resilience assessment approach discussed in this report focuses mostly on diversification, by considering both DoD and commercial assets, the framework can also capture the impact of the other methodologies on mission resilience (and performance) when the effects

---

[5] See discussions in Shaw, Miller, and Hastings (2001) on Type 1 adaptability metrics.

[6] Joint Publication (JP) 3-14, *Space Operations*, Joint Chiefs of Staff, October 26, 2020, p. I-8.

[7] Two examples include Office of the Assistant Secretary of Defense for Homeland Defense and Global Security, *Space Domain Mission Assurance: A Resilience Taxonomy*, September 2015; and Burch, 2019.

[8] JP 3-14, 2020, p. I-8.

[9] JP 3-14, 2020, pp. I-9 – I-10.

of the methods can be captured in the simulation. As an example, when disaggregation is used to improve resilience, the effectiveness of the method can be assessed by simulating the disaggregated system in a physics-based model. This is also true for the other suggested resilience methodologies as long as the effect of the method can be properly simulated. For our purposes, our implementation of a resilience assessment framework aims to estimate how key performance parameters (KPPs) for specific missions vary in response to system changes.

As an example, a KPP could be the revisit rate for satellites over a given area or target. We define a resilience measure by assessing how the revisit rate is affected by a degradation to the USSF satellites of a given architecture and then consider if and how commercial services included in the architecture can mitigate the impact of that degradation. This *buy back*, or measure of resilience provided by commercial services, could be in the form of how well the degraded revisit rate improves as more commercial services are added. Additional considerations are involved, including the quality of the measurements and the trustworthiness of the data, that will affect the value of the buy back.

The PAF-developed resilience assessment framework uses a combination of physics-based models and post-processing systems analysis to perform the assessment. A high-level view of the framework is shown in Figure 1.1. The framework contains four broad components: (1) identify need, (2) estimate USSF capabilities, (3) estimate commercial capabilities, and (4) calculate resilience measure. In the remainder of this section, we describe each step in the framework and conclude with a discussion of the advantages and limitations of the described framework.

Figure 1.1. Resilience Assessment Framework

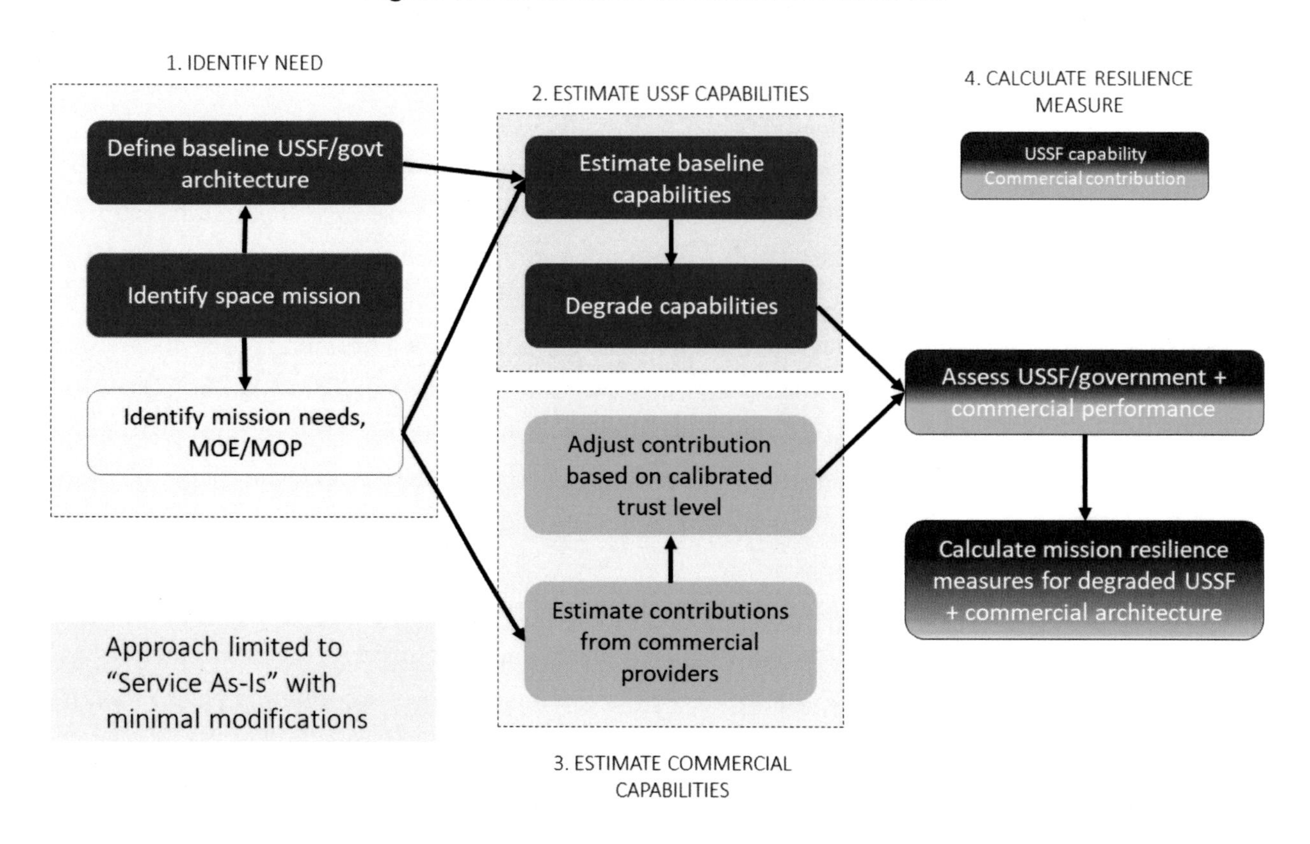

NOTE: govt = government; MOE = measure of effectiveness; MOP = measure of performance.

Identify Need

The first step in the framework (left-hand box) involves identifying the USSF mission system's architecture, as well as identifying the various measures of performance (MOPs) associated with the mission. This step includes defining the baseline government architecture and potential commercial service or architectural augmentations.

Estimate USSF Capabilities

The second step in the framework (top center box) involves estimating the performance of the baseline USSF (or DoD) architecture. Implementation of this step (as well as the third step) can be decomposed into two main subcomponents: the *physics-based modeling* and the *post-processing systems analysis* for resilience assessments. We describe these subcomponents here only briefly; a more in-depth discussion of them follows in Chapter 2.

The goal of the physics-based modeling subcomponent is to produce a simplified representation of the USSF mission system that can be (1) used to estimate mission-specific MOPs and (2) easily modified to simulate the effects of specific mission system interventions (e.g., asset degradation). We used the Advanced Framework for Simulation, Integration and Modeling (AFSIM) for this modeling task.[10]

The systems analysis subcomponent applies a series of computational post-processing steps to the output of the modeling step. The goal is to simulate a portfolio of interventions (e.g., stresses or changes) to the mission system using the physics-based output and then to estimate the mission system's performance in response to those simulated interventions.

Estimate Commercial Capabilities

This step introduces the contributions from the selected commercial provider(s) system and estimates the impacts of their contributions on mission MOPs. The commercial mission system undergoes physics-based modeling like the baseline system. The structure of the systems analysis workflow described in the previous section allows us to combine baseline and commercial contributions via a simple computational augmentation of the baseline mission model (see Chapter 2 for methodology details).

The introduction of third-party commercial contributions of information to DoD missions raises questions of trustworthiness. Our resilience assessment framework includes a subframework that allows analysts to explicitly account for the trustworthiness of information. In this subframework, the model conducts and fuses trust evaluations on *both* the information sources (the commercial organizations) and the individual information artifacts. We discuss a hierarchical approach that we explored for integrating signals of trustworthiness into resilience assessments in Chapter 3.

Calculate Resilience Measure

The fourth step in the resilience framework (rightmost box) involves calculating the mission resilience of various instances of the mission system under study. This includes an evaluation of the performance of the baseline USSF-only mission system and the USSF + Commercial mission system (i.e., the USSF mission system augmented with commercial capabilities) using the modeling and systems analysis workflow introduced above. We then evaluate how both architectures respond to degradations of varying intensity. These estimated system responses are the primary data used to inform statements about mission system resilience.

Organization of This Report

As indicated previously, the next two chapters describe the methodology developed for the post-processing systems analysis approach (Chapter 2) and the trust assessment (Chapter 3). In

---

[10] Peter D. Clive, Jeffrey A. Johnson, Michael J. Moss, James M. Zeh, Brian M. Birkmire, and Douglas D. Hodson, "Advanced Framework for Simulation, Integration and Modeling (AFSIM)," in *Proceedings of the International Conference on Scientific Computing (CSC)*, Steering Committee of the World Congress in Computer Science, Computer Engineering and Applied Computing (WorldComp), 2015.

Chapter 4, we illustrate how the framework could be implemented using two USSF missions as examples:

- *Tactical intelligence, surveillance, and reconnaissance (T-ISR)*. The goal of this mission is to apply sensor assets in orbit to develop situational awareness about a portfolio of operationally relevant terrestrial locations.
- *Data transmit and receive network (DTRN)*. The DTRN mission is a subcomponent of the USSF's Ground Enterprise Next (GEN) effort. The DTRN is intended to serve as a reliable data communication infrastructure connecting data-acquiring space assets with ground assets that receive, relay, process, and/or store acquired data.

Finally, in Chapter 5, we briefly discuss limitations of the framework and identify future steps that can be taken to enhance the robustness of the framework.

Chapter 2. Estimating Mission Capabilities and Resilience

To determine how resilient a satellite constellation is under different assumptions, we developed a resilience approach that performs a quantitative evaluation of how tractable models of USSF mission systems respond to changes in the systems' structure. The aim is to derive insight into how mission performance responds to changes in the system being used to conduct the mission. In the context of this study, we focus on two structural changes: (1) the augmentation of USSF capabilities with mission-relevant commercial offerings; and (2) the random degradation of the mission system's configuration. The random degradation is intended to represent a simplified simulation of operational "wear and tear" that can affect the availability of mission system components. Further work could also implement targeted interventions to a mission system to simulate adversarial attacks on USSF mission–relevant infrastructure.

The remainder of this chapter describes the steps entailed in the *physics-based modeling* and *post-processing systems analysis* framework subcomponents introduced in the previous chapter. We can further disaggregate these subcomponents into three key computational steps as follows:

- physics-based modeling
  - *System representation*: produce a tractable representation and simulation of the mission system under evaluation. We apply physics-based modeling and simulation tools for this step. We only need to do this once for the baseline and commercial architectures.
- post-processing systems analysis
  - *System evaluation*: extract MOPs for different configurations of the mission system under study. This is implemented as post-processing computations on the output of the physics-based model representing the mission. There is typically more than one MOP of interest for a mission under evaluation.
  - *Degradation analyses*: subject the system representation to stylized degradation and evaluate the system's response. This is also implemented as a post-processing computation on the output of the physics-based model. This approach assesses the degraded performance of the architecture without having to run a physics-based model for each degraded architecture.

Various parts of this assessment framework draw on preexisting ideas on the evaluation of the fragility of complex systems. For example, Welburn et al. (2020) uses simulations on a network representation of the U.S. economy to evaluate systemic risk in the economy.[11] And

[11] Jonathan William Welburn, Aaron Strong, Florentine Eloundou Nekoul, Justin Grana, Krystyna Marcinek, Osonde A. Osoba, Nirabh Koirala, and Claude Messan Setodji, *Systemic Risk in the Broad Economy: Interfirm Networks and Shocks in the U.S. Economy*, RAND Corporation, RR-4185-RC, 2020.

Shaw (1999) formalizes in detail how space mission systems are generalized information transmission networks whose systemic properties can be characterized quantitatively.[12]

System Representation

The first step of the physics-based modeling is to produce a tractable simplified representation and simulation of the mission system under evaluation. We need a workflow that is capable of computationally representing the key aspects of a given space mission while abstracting away details that are not relevant or influential for evaluating KPPs for the mission. The approach needs to strike a balance between completely *faithful* (but unwieldy) representations of the space mission versus careful abstractions of the mission that still allow us to make useful performance evaluations. We have some flexibility in implementing this representation. For this study we experimented with *homegrown* simulations (in Python), as well as simulations based on AFSIM (a mission-level simulation environment that can be used to model space architectures).[13] We found the AFSIM tool to be a better option because of the richness of its representations of the space missions that we examined.

Physics-Based Model Using AFSIM

Physics-based modeling can provide, as needed, foundational data for estimating the capabilities and resilience for both the USSF and the commercial architectures. For the purpose of resilience assessment, the main output from a given AFSIM simulation run is the database of observation or accessibility events that the model identifies between assets in a given space constellation and a portfolio of ground targets. The ground targets can, for example, represent terrestrial points of interest (POIs) for a T-ISR mission or ground antennae location for a DTRN mission. The resulting event database from AFSIM is a robust-enough simulation of how a given asset configuration performs a mission under study. This event-based dataset becomes the primary input for subsequent post-processing analyses that estimate the mission's KPPs and produce resilience assessments.

As an example, we consider the average gap times to image multiple targets of interest as a KPP for the T-ISR mission. For the baseline USSF-only architecture, the approach involves modeling the architecture using a physics-based model and generating an output file that includes a *time series* of observation opportunities or events for a portfolio of terrestrial nodes representing target or antenna locations (see Table 2.1 for a sample time series).[14] We then

[12]Graeme Barrington Shaw, *The Generalized Information Network Analysis Methodology for Distributed Satellite Systems*, dissertation, Massachusetts Institute of Technology, 1999.

[13] Clive et al., 2015.

[14] We used AFSIM in developing this framework, but this step can be reproduced with any government or commercial tool that can reproduce similar event-based data, such as the Ansys Systems Tool Kit (STK).

model the ability of the commercial architecture of interest to perform the same task as the baseline USSF-only architecture. The modeling can also impose asset constraints (e.g., the resolution of the image[15] or antennae's supported communication bands) as part of the requirements specified by the USSF for a given mission.

We used AFSIM to estimate the accesses for both example the USSF and commercial architectures used in our analyses. In future versions of this modeling workflow, commercial providers may also deliver time series of its observation capability to the USSF. Ideally the information would be presented in a standard format as a function of time and for requested ground nodes of interest. This information would alleviate the need for the USSF to model the providers' constellation. Analyses would also need to account for scheduling and availability constraints on commercial constellations that may support the USSF.

Table 2.1. Sample Observation Event Data for the DTRN Mission from the AFSIM Model

Time (seconds)	Ground Asset ID	Space Asset ID	Distance (km)	Elevation (degrees)
280	Gov_6	LEO_CONSTELLATION_19_18_1	581.47768	75.4109336
290	Gov_6	LEO_CONSTELLATION_19_18_1	581.47768	75.4109336
330	Gov_6	LEO_CONSTELLATION_19_18_1	581.47768	75.4109336
1,030	Com_14	LEO_CONSTELLATION_19_8_0	578.16196	75.6163696
1,040	Com_14	LEO_CONSTELLATION_19_8_0	578.16196	75.6163696
1,060	Com_14	LEO_CONSTELLATION_19_8_0	578.16196	75.6163696
1,070	Com_14	LEO_CONSTELLATION_19_8_0	578.16196	75.6163696
2,310	Com_9	LEO_CONSTELLATION_19_7_1	567.80312	74.2641004

NOTE: Com = commercial; Gov = government; LEO = low earth orbit.

A representation of the sample time-series output is shown in Figure 2.1, which depicts results of a notional USSF-only architecture combined with commercial data. The dots on the globe represents the *ground nodes of interest*, i.e., where potential targets are located and are captured on the vertical axis of the plot. The horizontal axis represents time, with zero as a reference or starting time. The USSF-only architecture consists of satellites A, B, and C, while the commercial architecture is depicted by satellites D and E. The node diagram captures which satellites provide a measurement at what time. For example, at time 0, node 6 is accessed by satellites E and C (represented by E(0) and C(0) in the upper left cell).

[15] Lower resolution images can also be valuable if it can provide the USSF architecture with a cue, e.g., activities of interest at location x that should be prioritized in scheduling; or, no activities or targets at location x so imaging can be delayed. This type of cueing would help the USSF improve the use of its limited resources.

Figure 2.1. Illustration of Observation Events Recorded in a Sample Output from the Physics-Based Model

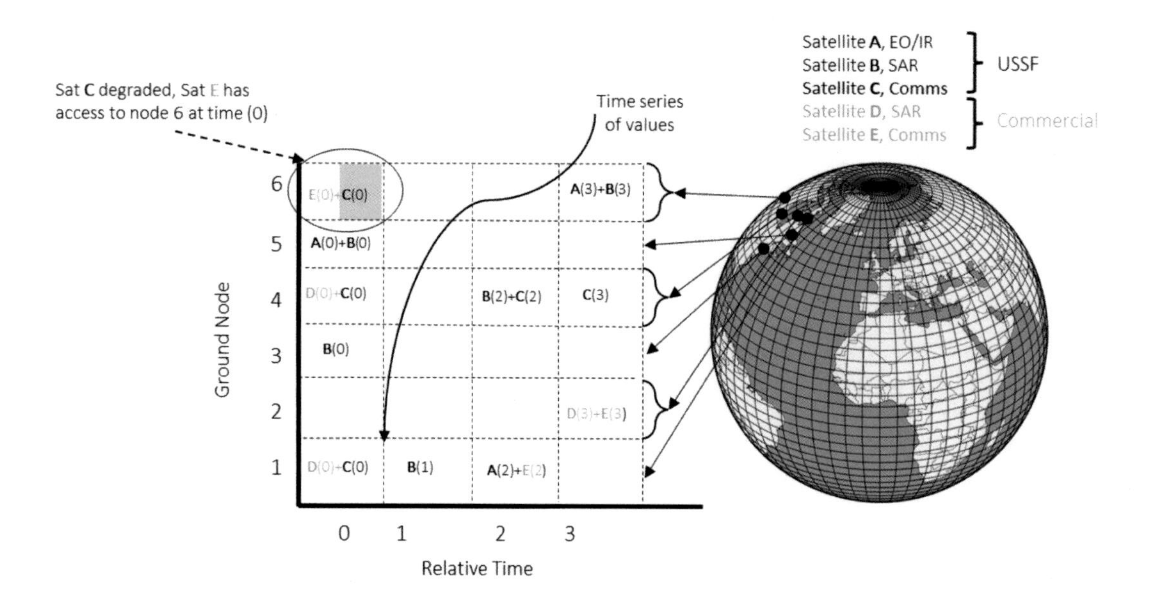

NOTE: Comms = communications; EO/IR = electro-optical/infrared; SAR = synthetic aperture radar; Sat = satellite.

A Dynamic Network Interpretation of the Model

The physics-based model of space mission configurations admits an alternate equivalent (and general) description using the language of network analysis.

Space missions (especially missions facing the Earth) typically involve the transmission and reception of electromagnetic waves between assets in orbit and assets and points on earth. For example, the DTRN mission involves the transmission and reception of radio waves for communicating digital and analog data. The T-ISR mission involves the transmission and reception of EO/IR and radar electromagnetic radiation between satellites and terrestrial points. We can model the space and terrestrial assets as *nodes* in a notional network. The *edges* of that network can represent opportunities for electromagnetic transmission and reception among the nodes (e.g., when an EO/IR satellite is within line of sight of a terrestrial POI).[16] The nodes in space and on earth are in continuous relative motion. As a result, the access edges fade in and out of existence over time.

[16] For example, these edges may be unrealized or just virtual if the satellite did not actually capture a POI that was in line of sight because of task scheduling, maintenance, or duty cycle constraints.

This description highlights the fact that the nodes and time-varying edges form a *dynamic network.*[17] The observation event database that the physics-based model produces (e.g., as depicted in Figure 2.1) is thus simply a dynamic network representation for the mission being modeled. The columns in the figure (the y-axis) specify adjacency lists for network over time (the x-axis).

This dynamic network frame also makes it easy to see key properties of the way we have set up the assessment framework. One key property is that the simulation is *highly modular*. Adding and subtracting satellites or ground POIs is as easy and trivial as adding or subtracting nodes from the network. We do not have to rerun expensive physics-based simulations for the entire augmented constellation, just for the new nodes. This approach significantly reduces the needed number of physics-based models when assessing the impact of a degraded architecture. For example, we can easily remove a satellite from the network and eliminate the contribution of that satellite to the mission. This capability allows us to assess a large number of degraded architectures without the need to rerun the physics-based models.

System Evaluation

A mission system's *configuration* (i.e., the collection of all space and ground assets in play) determines how well the mission can be performed. The observation event database from the physics-based model gives us a flexible means for estimating many relevant MOPs for a given mission system. It also gives us the means to evaluate how (limited) changes to the mission system affect MOPs. What we refer to as *resilience analysis* is the process of evaluating the mission's MOP response to structural changes in mission systems. The first consideration for system evaluation is specifying relevant MOPs. Examples of MOPs that we used include the following:

- for the T-ISR mission: total time needed to acquire all portfolio targets and the revisit rate for coverage of all target locations
- for the DTRN mission: average utilization, total connection time, and average satellite revisit times.

We developed a *post-processing systems analysis* workflow to estimate MOPs for selected missions by processing the event database produced by the physics-based model. We use this process to derive estimates of MOPs for various configurations of mission systems. Variations in these MOP estimates inform our conclusions on resilience. Each MOP calculation requires different computations on the event database, so we do not describe each MOP calculation here. In general, the database is flexible enough to accommodate many of the relevant MOPs that we

[17] Kathleen M. Carley, Jana Diesner, Jeffrey Reminga, and Maksim Tsvetovat, "Toward an Interoperable Dynamic Network Analysis Toolkit," *Decision Support Systems*, Vol. 43, No. 4, August 2007.

selected for the missions because, for both missions, performance relies on the exchange of information between assets on earth and in space. The event database captures all such exchange events (or just opportunities for such exchanges). We typically estimate mission MOPs under the mission system scenarios listed in Table 2.2. The process for performing simulated degradation for the degrade scenarios is described in the next section.

Table 2.2. Sample Analytic Scenarios for Evaluation

Degradation Level	USSF-Only Architecture	USSF + Commercial Architecture
Undegraded	Baseline case with no commercial support	Baseline case with commercial support (hypothetical)
Degraded (varying intensity)	USSF-only simulated stress tests for resilience evaluation	Simulated stress tests for resilience evaluation on USSF + Commercial system

Degradation Analyses

The systems analysis tool applies simulated degradation to a given mission architecture and evaluates the impact of degrades on estimated mission performance. Simulated degrades work by simulating the disabling of one or more nodes (satellites or ground POIs) and observing the impact on the MOP. The degradation can be done by time, for select satellite(s), by location, or any combination of the three. For example, ground node 2 may become unobservable or inaccessible because of jamming. The degradation can also be random or targeted. The systems analysis tool performs the degradation and assesses the impact on the MOP. The effects of these degrades can also be measured and compared against required performance levels.

The baseline mission representation is the output of the AFSIM modeling as described in the discussion above. Once the baseline capabilities are defined, we post-process the event time-series database generated by the physics-based model and perform the systems evaluation process for different instantiations of a degraded architecture. Repeating this process for degrade scenarios at different intensities of degradation provides quantitative information for evaluating the mission resilience of systems. We compared this performance information across all four system scenarios shown in Table 2.2

Mission performance is expected to decline as the intensity of degradation increases. But the rates of performance decline (one measure of resilience) and ability to exceed prespecified performance thresholds will differ based on the configuration of the mission system. These variations are what we aimed to report for our case studies.

Chapter 3. Trust Assessment

Leveraging commercial companies to provide T-ISR data poses a unique trust challenge. DoD has traditionally relied on a sensor-to-data supply chain that is fully DoD developed, maintained, and controlled. Using commercial imagery data from non-DoD owned assets is a departure from this norm and requires a trust assessment, especially concerning data veracity. For our purposes, we define *trust* simply as a measure of how confident we are that a supplied piece of information conforms to the truth.

Data Supply Chain for a Commercial Partner

A notional overhead image passes through many proverbial hands prior to delivery to the USSF, as illustrated by the notional path shown in Figure 3.1. A satellite will image a region of the earth via its EO sensor. The raw data of the image are transmitted via radio waves to a ground station. At the ground station, the data are stored on servers, awaiting processing, which may be performed in the cloud on at an offsite location. Image processing occurs in a well-defined stage-by-stage process. To begin, raw data (Level 0) is converted into an image (Level 1) and distortions are corrected (both radiometric and geographic).[18] At the next stage (Level 2), known geographical information, such as elevation data, is incorporated into the image. Level 3 applies additional correction to ensure the image matches known cartographic grids and scales of the area. Level 4 adds further information to the image, fusing data from models, other sensor sources, and multiple measurements.

The USSF may request information at a given processing level (L0–L4), depending on its information needs. More highly processed images provide more useful information while carrying more risk: Such images have been manipulated more often. Figure 3.1 identifies points where vulnerabilities may exist. The first concern is whether the image was actually taken by the alleged satellite. Second, because the imagery data sit on various servers as processing jobs are pending, there is the concern that unauthorized changes have been performed on the data during this time. Third, the data undergoes intentional manipulation during image processing. Can we audit the underlying processing code to ensure malicious changes are not occurring? Finally, once the data are sent to USSF servers, can we verify that the data were uploaded by the company?

[18] Data processing levels reflect those used by Earthdata, "Data Processing Levels," National Aeronautics and Space Administration, last updated July 13, 2021.

Figure 3.1. Notional Data Supply Chain to the USSF

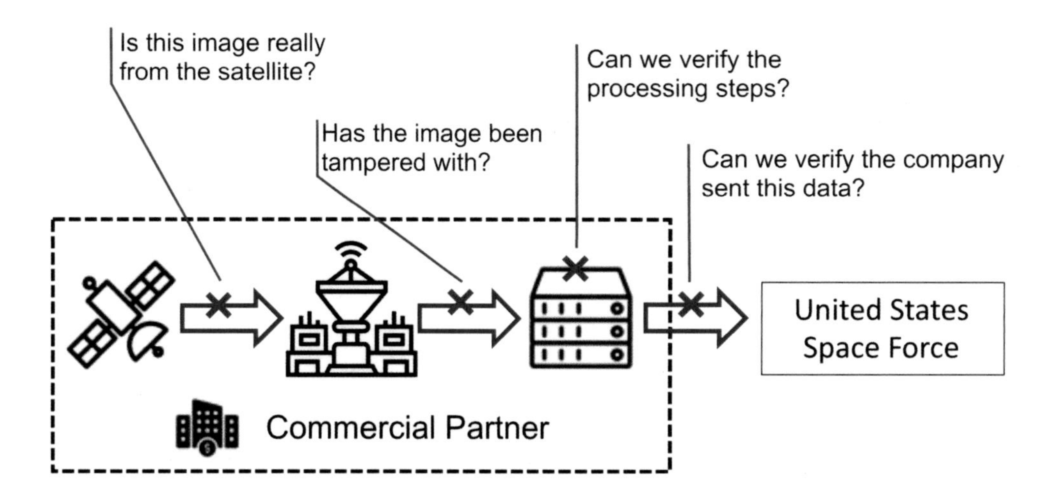

Trust: A Layered Approach

Numerous papers have been written on the subject of trust, and various approaches have been developed for assessing trust and fusing information with varying trust levels.[19] We propose the use of a *layered approach* for assessing the trust associated with information from commercial sources. The layers involve different methodologies, each of which independently yields a trust value or score. The number and context of the layers depend on the mission they are being applied to, such as T-ISR, satellite communications, or others. We do not intend for the layers to always be used or needed; instead, we suggest using as many layers as possible given the existing conditions, e.g., number of sources, data availability. We use one of our case-study missions, T-ISR, to illustrate the approach and provide example results. An example of this layered approach, shown in Figure 3.2, contains five layers:

- integrate multiple sources
- perform an a priori assessment
- use a "fingerprinting" approach
- use encryption technology
- leverage physics-based modeling.

[19] David A. Nevell, Simon R. Maskell, Paul R. Horridge, and Hayleigh L. Barnett, "Fusion of Data from Sources with Different Levels of Trust," *2010 13th International Conference on Information Fusion*, 2010; Sean Kinser, Pete de Graaf, Matthew Stein, Frank Hughey, Rob Roller, David Voss, and Amanda Salmoiraghi, "Scoring Trust Across Hybrid-Space: A Quantitative Framework Designed to Calculate Cybersecurity Ratings, Measures, and Metrics to Inform a Trust Score," Small Satellite Conference, 2020.

We then use an evidence-based theory approach, namely the Dempster-Shafer theory (DST),[20] to combine the results of each layer to obtain an overall trust score. We describe these layers in the sections that follow, including a more in-depth discussion of the "fingerprinting" approach.

Figure 3.2. Example of Layered Approach to Trust Assessment

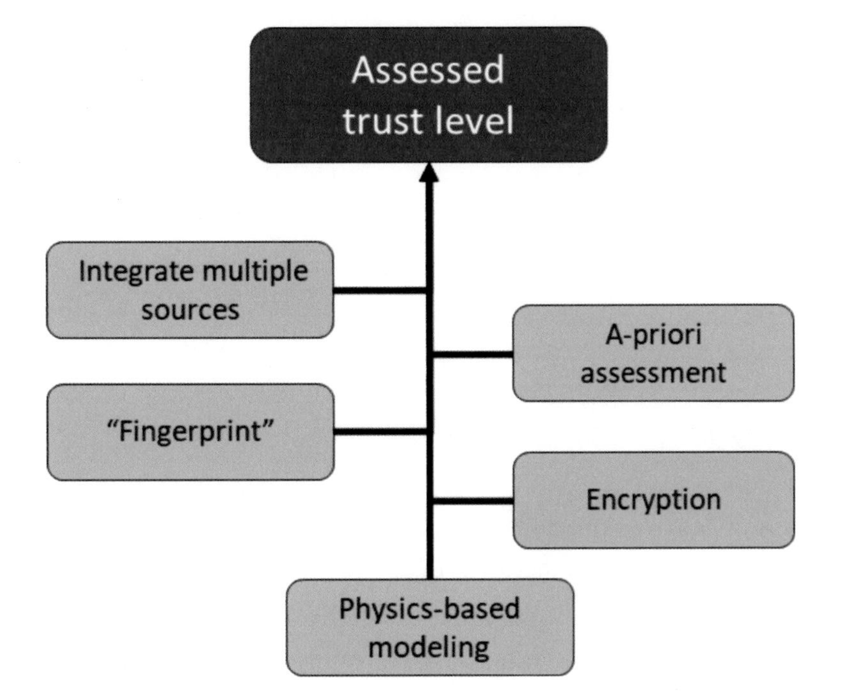

Integrate Multiple Sources

Integrating multiple sources is a well-known concept that has been applied to sources with varying levels of trust.[21] For the T-ISR mission, it involves leveraging information about a target from different sensors, possibly of different phenomenology (e.g., passive and active). The information from the independent sources can be used in different ways; in a tracking case, for example, it might be desirable to integrate the tracking information to reduce the error associated with a track. However, for the purposes of assessing trust, the information from the different sources would be compared with each other to ensure they are consistent. We consider two scenarios and approaches by

- comparing the information from two or more independent commercial providers and determining a trust score based on the consistency between the different sources

---

[20] Arthur P. Dempster, "A Generalization of Bayesian Inference," *Journal of the Royal Statistical Society: Series B (Methodological)*, Vol. 30, No. 2, 1968; Glenn Shafer, *A Mathematical Theory of Evidence*, Princeton University Press, 1976.

[21] Nevell et al., 2010.

- verifying the information from a commercial provider by comparing a subset of the data with a higher trust government (or similar) source.

The outcome from the analysis could be to subject the information from the different sources to these additional verification steps to decide which one is more trustworthy (if they are inconsistent with each other). Otherwise, if the data are inconsistent and no other assessment methods are available, then all the involved sources would need to be considered untrustworthy until further assessment.

A Priori Assessment

An *a priori assessment* involves, as the name implies, an evaluation of the trust prior to the generation or use of the information. This approach is widely applied by various organizations within and outside the government—one example being to assess the security posture of cyber systems.[22] We consider the use of the Cybersecurity Maturity Model Certification (CMMC) framework, developed by the Office of the Under Secretary of Defense for Acquisition and Sustainment, as the a priori cybersecurity-related trust assessment tool. It scores a provider from a scale of 1 to 5, with 5 being the highest maturity (or trust for our purposes). Although the CMMC approach offers some reassurance about the protection of data, it does not necessarily provide an assessment of the contractor's ability to protect these data consistently, such as whether the provider ensures daily adherence to processes in place or zero-day threats.

Another approach assessing the trust associated with commercial providers uses an Architecture Score Index (ASI) as a metric of the overall trust in the data received.[23] The ASI framework uses a three-phased approach and combines the results to calculate the ASI score. The first phase uses the CMMC as a baseline for assessing the organization; phase 2 considers daily cyber hygiene activities, such as patch installation; and phase 3 considers incident response in terms of mean time to detect and to resolve incidents. The outcomes of each phase for a system are quantified, weighted, normalized, and then combined to produce the ASI or trust score.[24] We consider the ASI methodology to be a significant improvement over CMMC because the daily performance of the contractor is considered and accounted for in the score, and we therefore recommend its use, when possible, to develop the a priori–related trust score.

---

[22] Steven Geraldo and Randy Woods, "CISA's Cybersecurity Evaluation Tool (CSET)," presentation at the Chemical Security Seminars Virtual Expo, December 16, 2020.

[23] Kinser et al., 2020.

[24] Kinser et al., 2020.

Fingerprinting

A key feature of a layered trust approach is to independently verify (or at least ascertain, with some probabilistic confidence) various aspects of the data supply chain. *Fingerprinting* is an image forensics technique to verify if a given image was taken by a purported sensor.

Fingerprinting relies on extracting data encoded within an image and comparing that data with known information about the sensor. There are three common fingerprinting techniques, which we discuss in order of increasing strength. The first is reading the metadata of an image. Camera systems typically embed information onto images using an exchangeable image file format, or EXIF header. Verifying this content may offer some amount of trust, but this technique is extremely brittle. The information exists on the image as plain text and can be modified by anyone, including an adversary. Second, every sensor leaves a unique but fragile fingerprint on an image due to sensor-level differences in light capture. We consider this technique to be of medium strength. Third, hash verification provides strong evidence that a given image is legitimate. Hashing algorithms analyze the byte contents of an image to compute a unique, image-specific text string.[25] A calculated hash can be compared with its known, trusted value to determine if it has been tampered with in any way. This technique is very strong but poses a practical problem: How do we establish the trusted hash of an image with which to perform this comparison? This will be explored in the next section.

A U.S. patent identified a method to exploit sensor-specific differences and verify if images were taken from a sensor.[26] The subsequent paper by the patent authors provides additional details about this technique.[27] Sensor-specific pattern noise is present on all EO sensors, and the largest contributor is the photon response nonuniformity (PRNU) present in sensor wafers. EO sensors consist of many individual photonic sensors that capture light from a scene. Each of these photonic sensors responds to light slightly differently. Collected photons are converted into voltages, which are measured and compiled across the entire sensor to form a raw digital image.

PRNU presents itself in digital images as high-frequency noise. To extract the PRNU from a given image, the image undergoes wavelet denoising to remove high-frequency noise. This noise is recovered by subtracting the denoised image from the original. Figure 3.3 visualizes the PRNU that has been extracted from an image. In the right image, an outline of the woman and the building in the background are visible. This illustrates a practical issue with the extraction process: Content-specific information is extracted in the PRNU calculation. The developers of

---

[25] A hash is mathematically not unique by design. Hash collisions occur when two disparate files share the same hash. However, the probability of this occurring in commonly used hash functions are practically zero. For all intents and purposes, we assume a safe hash function is used to generate unique hashes (e.g., SHA-256).

[26] Jessica Fridrich, Miroslav Goljan, and Jan Lukáš, *Method and Apparatus for Identifying an Imaging Device*, U.S. Patent 7,787,030 B2, August 31, 2010.

[27] Jan Lukáš, Jessica Fridrich, and Miroslav Goljan, "Digital Camera Identification from Sensor Pattern Noise," *IEEE Transactions on Information Forensics and Security*, Vol. 1, No. 2, June 2006.

the technique describe creating a PRNU fingerprint for a sensor by averaging the PRNU calculations from at least 50 images.[28]

To test if an unknown image was taken from a specific sensor, the PRNU of the unknown image is compared with a PRNU fingerprint via correlation analysis. If the correlation coefficient exceeds a preestablished threshold, the unknown image is verified to be taken from the sensor.

Figure 3.3. Example of PRNU Extraction

SOURCE: U.S. Navy photo by Victor Jorgensen (left) and authors' PRNU analysis (right).

A major issue with this fingerprinting approach is that it is vulnerable to some camera systems, processing techniques, or a resourceful adversary. The original patent admits that a process known as flat-field correction removes PRNU noise.[29] Advanced camera systems are capable of removing PRNU from each image by measuring the PRNU noise. Furthermore, a subsequent patent confirms this process as a practical anonymization technique to prevent sensor fingerprinting.[30]

We tested the PRNU fingerprinting technique on open-source satellite data. We selected xView2, a large satellite dataset provided by the Defense Innovation Unit for an artificial intelligence (AI) object-recognition challenge, for analysis.[31] This dataset consists of images taken by three satellites: Geoeye01 (2,385 images), Worldview02 (4,409 images), and

---

[28] Lukáš, Fridrich, and Goljan, 2006.

[29] Flat-field correction is effectively the first half of the PRNU extraction process: denoising the image. See Fridrich, Goljan, and Lukáš, 2010.

[30] Ahmet Emir Dirik and Ahmet Karakucuk, *Anonymization System and Method for Digital Images*, patent application, World Intellectual Property Organization International Bureau, International Publication Number WO2014/163597A2, October 9, 2014.

[31] Defense Innovation Unit, "xView Challenge Series," webpage, undated.

Worldview03 (670 images). It is unclear which processing level the xView2 dataset is, but we estimate it to be Level 2 or 3 because of the high amount of processing in the images (i.e., orthorectification) to make it suitable for a public AI challenge.

Testing PRNU on Satellite Data

To test PRNU on satellite data, we used images only from Geoeye01 and Worldview02 because their images dominated the dataset. We calculated sensor fingerprints for each satellite using 500 images for each and tested the fingerprinting correlation technique with the remainder of images from each sensor. Figure 3.4 gives a visual example of the Geoeye01 sensor's average PRNU (brightened for visibility). As the figure shows, the noise pattern does not appear to provide a good visual discriminant for fingerprinting, and as it turns out, this is true algorithmically as well (see Figure 3.5).

Figure 3.4. Example of PRNU Analysis of Geoeye01 Sensor Data

SOURCE: Features information from xView2 database accessed from Defense Innovation Unit, undated.
NOTE: The noise fingerprints are hard to meaningfully distinguish or classify visually even though we brightened the image for clarity.

Sensor predictions taken from the maximally correlated sensor resulted in an accuracy of 56 percent. We generated receiver operating curves to test the effects of correlation thresholds on accuracy. The area under the curves provides a summary of the segregation ability of using correlation values to distinguish between two sensors. The values were 0.46 and 0.59 (depending on the sensor fingerprint for which we were calculating the correlation values), which indicated poor performance. Upon further investigation, we discovered that correlation values do not offer much segregation. Figure 3.5 shows that the distributions of correlations overlap, making it a poor differentiator.

We abandoned the correlation-based approach and tested a support vector machine (SVM) to determine if a machine learning (ML) classifier can distinguish the PRNU calculations between these two sensors. We trained the SVM on the 500 PRNU calculations from each sensor and assessed its performance on the remainder of the dataset. The SVM performed with 51–66 percent accuracy, depending on the kernel type (Table 3.1).

Figure 3.5. Distributions of Correlations Indicate No Segregation

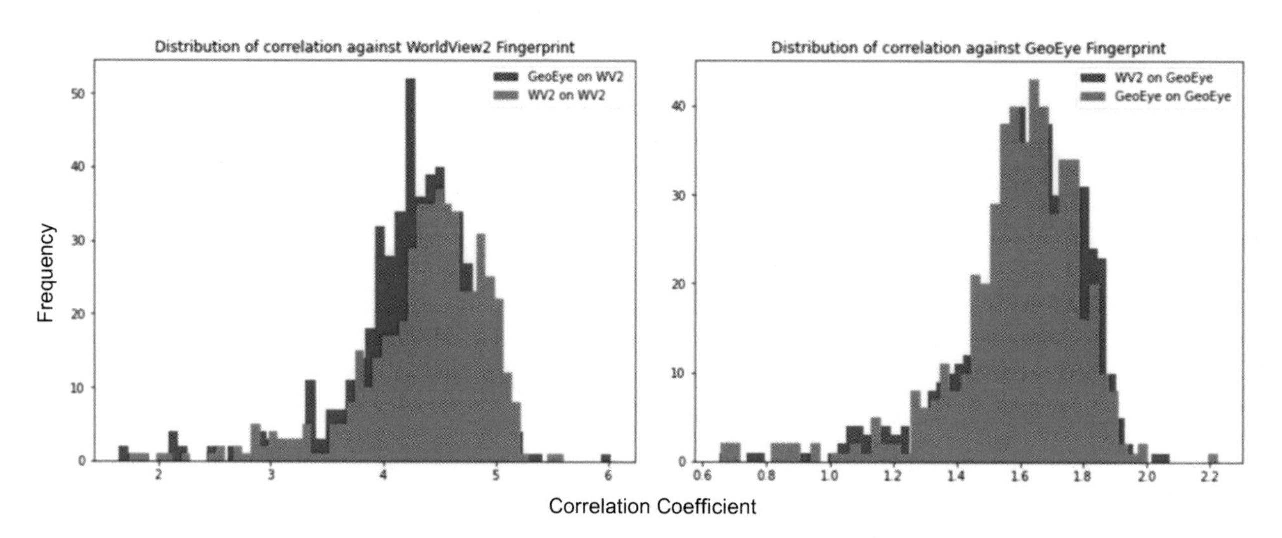

NOTE: GeoEye = Geoeye01 satellite; WV2 = Worldview02 satellite.

Table 3.1. Test Set Accuracy for Support Vector Machine Sensor Classifier

Kernel Type	Test Set Accuracy
Linear	51%
Third degree polynomial	52%
Radial basis function	66%

Finally, we tested the ability of a neural network to perform PRNU segregation. For computational simplicity, we reduced the PRNU calculations down from the original image shape of 1024 x 1024 pixels to 128 x 128 pixels. This prevented the network from being too large. We used a convolutional neural network (CNN) architecture to output the probability of the image belonging to each sensor.[32] Prediction performance on both the training and test set capped around 53 percent and demonstrated no improvement after 1,000 training epochs.

We note, however, that PRNU fingerprinting may not be a viable method for highly processed satellite data. According to a discussion with a commercial satellite imagery provider, flat-field correction is a standard calculation during Level 1 processing of satellite images.[33] This likely explains why we were unable to use PRNU noise successfully. Our investigation revealed that Level 0 satellite data is difficult to obtain and that most commercial providers do not typically offer this product.[34] One of the largest open-source satellite imagery data provider, the National Aeronautics and Space Administration (NASA), offers some Level 0 products. However, we were unable to work with any of these data in our limited analysis because the file format specifications required special handling. We leave testing the PRNU technique on Level 0 data for future work. Finally, we note that other researchers have been unable to use PRNU to perform sensor identification for a variety of common commercial camera systems.[35]

An additional vulnerability we discovered was that PRNU noise is specific to the sensor orientation. The averaging process to obtain a sensor fingerprint requires that all images be oriented in the same direction to preserve the necessary information. If any of the sensor images are rotated, which is often the case in satellite imagery (e.g., to ensure north points up), the method will potentially fail. This rotation is common in higher image processing levels. We are uncertain whether the xView2 images that we analyzed were oriented uniformly across sensors.

Neural Network Fingerprinting

The PRNU-based fingerprinting approach may be relatively weak compared with newer approaches such as *noiseprint*, a CNN-based fingerprint identification.[36] Supplementing PRNU with noiseprint demonstrates remarkable segregation abilities with area-under-the-curve values of 0.90 or better.[37]

Taking inspiration from these techniques, we tested a neural network's ability to ingest satellite images (not PRNU calculations) and perform sensor identification. We expanded our

---

[32] CNN architecture is as follows: input of 128 x 128 x 3, CNN32, CNN64, CNN64, fully connected (FC) 8, FC2, with a 2 x 2 MaxPool operation following each CNN.

[33] Commercial satellite imagery provider, discussion with the authors, March 2021.

[34] However, some of the feedback we received from a commercial satellite imagery provider indicates that it would be possible for the government to receive Level 0 data if providers were under contract to do so (commercial satellite imagery provider, discussion with the authors, March 2021).

[35] Eleni Athanasiadou, Zeno Geradts, and Erwin Van Eijk, "Camera Recognition with Deep Learning," *Forensic Sciences Research*, Vol. 3, No. 3, July 2018.

[36] Davide Cozzolino and Luisa Verdoliva, "Noiseprint: A CNN-Based Camera Model Fingerprint," *IEEE Transaction on Information Forensics and Security*, Vol. 15, No. 1, 2020.

[37] Davide Cozzolino, Francesco Marra, Diego Gragnaniello, Giovanni Poggi, and Luisa Verdoliva, "Combining PRNU and Noiseprint for Robust and Efficient Device Source Identification," *EURASIP Journal on Information Security*, 2020.

dataset to include a total of four sensor types: Geoeye01 and Worldview02 from the xView2 image database, ALOS AVNIR-2 (a Level 1 image from NASA's Alaska Satellite Facility),[38] and deepfake satellite imagery.[39] *Deepfakes* are AI-generated data that include images, videos, and audio. These falsified data are convincing to a human and difficult to distinguish from authentic data (by design). We incorporated deepfake data to test a plausible scenario in which an adversary has somehow intercepted the data supply chain and is inserting fake imagery. The xView2 data consisted mostly of urban environments. The ALOS AVNIR-2 dataset consisted mostly of coastal images. The deepfake data consisted of falsified urban environments.

We designed the neural network to tackle a multiclass classification problem. The network architecture ingests an image of size 512 x 512 pixels processes it through several fully connected layers, and outputs a vector of length 4, representing the probability of belonging to each of the four classes.[40] Images larger than 512 x 512 pixels were scaled appropriately. We used 500 images from each sensor type as training data, and the remaining images were used to form a test set. We achieved 99-percent training accuracy and 90-percent test accuracy.

Our proof of concept with a CNN-only approach achieved superior sensor identification performance. Although we cannot be certain that the neural network was learning sensor-specific fingerprints versus image content–specific information, we demonstrated a practical and straightforward approach to perform sensor identification. More advanced techniques, such as specially designed CNN architectures to obtain sensor fingerprints, have established success rates.[41]

In summary, our research related to the fingerprinting approach indicates that although the PRNU option was not sufficiently reliable to support information assurance, using a CNN-based approach such as noiseprint appears very promising. Additional analysis using more datasets needs to be performed to further assess this neural network approach.

Encryption

While neural network sensor fingerprinting can provide confidence that a given image was taken by a purported satellite, that method only provides reassurance to one aspect of the data supply chain shown in Figure 3.1: How do we establish a chain of integrity between the satellite, ground station, processing servers, and the USSF?

---

[38] Alaska Satellite Facility, "ALOS AVNIR-2 Ortho Rectified Image Product," dataset, undated.

[39] Bo Zhao, Shaozeng Zhang, Chunxue Xu, Yifan Sun, and Chengbin Deng, "Deep Fake Geography? When Geospatial Data Encounter Artificial Intelligence," *Cartography and Geographic Information Science*, Vol. 48, No. 4, July 2021.

[40] Input (512 x 512 x 3) to FC32 to FC16 to FC8 to Output (FC4). Prediction taken by using the argmax of the FC4 output.

[41] CNN architecture is as follows: input of 128 x 128 x 3, CNN32, CNN64, CNN64, fully connected (FC) 8, FC2, with a 2 x 2 MaxPool operation following each CNN (Athanasiadou, Geradts, and Van Eijk, 2018).

We explore the use of image hashes to accomplish this task. As discussed earlier, hashes provide a reliably unique identifier for a given image. Hashes can be recorded every step along the data supply chain, from the satellite taking the image to the final product that the USSF receives. Critically, as the data undergo the sequence of processing steps, hashes of each step can be recorded. However, this raises another issue: How can the USSF ensure that it receives a series of trusted hashes? An adversary or compromised commercial entity can send maliciously altered data alongside its associated hash values.

The next solution lies in digital signatures. Encryption is a commonly known technique to render information unreadable to anyone without the proper decryption key. Private and public keys are popularly implemented in encryption: Anyone can encrypt information with a user's public key; however, only the user's private key can decrypt the information. Digital signatures act in reverse: A user's private key encrypts the data, and anyone can decrypt this information with the user's public key. For example, Alice sends Bob a message "Hello, world!" encrypted with her digital signature;[42] Bob can use Alice's public key to decrypt the signature and see the same "Hello, world!" message.[43] Because the two text strings match, Bob is assured that Alice indeed sent the message. If the message was not truly from Alice, then the decryption process would either fail or yield a non-matching message (likely gibberish).

As satellite data move along the supply chain, each node should receive the data, a digital signature, and the previous node's public key. This allows every node to independently verify whether the data they receive came from a trusted upstream participant in the supply chain.

A node can now independently verify the data from a node immediately preceding it. This raises two final problems. First, it may be possible that multiple upstream nodes have been compromised or that someone accidentally or intentionally has turned off signature-checking in an upstream node. How can a node independently verify the entire data supply chain leading up to the current data state? Second, nodes cannot verify whether the preceding node processed the data in an intended manner. A compromised node may perform malicious processing steps and create a valid digital signature on the malicious file.

One potential solution that simultaneously addresses both issues is to use a blockchain. The core blockchain principles are as follows: (1) all nodes participate in maintaining a shared immutable ledger, (2) ledgers record a chain of sequential events that any node can independently verify, and (3) nodes can write new events to the ledger in a secure fashion.

---

[42] An example of a digital signature, via RSA, a public-key cryptosystem:
A9ENUBB0BeCVn5pIS+IAKLYz0V/qerFQST7mY5RX4GSbDGws8Kl1Nvcb9qPofZ2OGHTPwVx6v8G+fvk/e0
kHC7TNhAgZpLa93444KKMJa1adKo2pYyC5QtnbaRgVgeYgDgQZmY1EySNUzmmhKE+EvLA6YNGoStic9H
O9vhEGm4k=

[43] An example of a public key generated by the authors:
MIGfMA0GCSqGSIb3DQEBAQUAA4GNADCBiQKBgQDCROQQiEX1nUdx/GpjAo6bkfleoDGLHxGG81V/n9
kwdzBkqYDl1zpIoFnu37/rzUnDkjHu3B36EZE9Xg9brjs8P13Pej+bONXyd3Gj/y6UyYWbeTtXbQDkjPr85kLhkK
X+Fk3/8UhYOUEV8Qx4U8BFIA57HCAXqVjPbQh8xUQNRQIDAQAB

Blockchain implementation may vary, especially in principle 3. Specific to this application, nodes perform each data-processing step in parallel. If most of these processing nodes agree on the end-result, the ledger is updated with that result's hash and the next processing step begins. Digital signature–checking is performed by each node to verify the data at each step. An adversary now needs to compromise more than 51 percent of the nodes in this data supply chain to create any modifications in the data. The USSF, upon receiving a commercial product, can use the ledger, a verifiable record of events, to confirm that the data were properly handled at each step in the supply chain.

Physics-Based Assessment

The final layer in this proposed trust assessment uses a physics-based model to verify the veracity of the information provided. Using the T-ISR example mission, a tool would check some known conditions, provided by independent sources, when the image was taken to ensure they are consistent with what is provided in the image. For example, satellite position, expected sun angle, and weather conditions would be compared with information contained in the image.

Fusing Layer Outputs into an Overall Trust Score

The final step in the trust estimation is to combine the various layers to calculate an overall trust score. The different layers described in the previous sections individually provide some quantitative or qualitative assessment of trust which need to be combined to obtain an overall trust score. We separate the CMMC assessment from the other layers because of the different characteristics and approach (i.e., CMMC represents the a priori and more qualitative assessment than the other layers).

We use evidence theory, specifically DST, for combining the assessment from the different layers (i.e., each layer is a piece of evidence).[44] DST is intended to address epistemic, or Type II, uncertainty, which is sometimes referred to as reducible because it can generally be reduced by additional empirical information, or evidence, at least in principle.[45] This theory is therefore a good application for calculating an overall trust score based on multiple inputs. The basic probability mass (bpm(x), or more simply $m(x)$), represents the degree to which the element x supports the claim represented by m. The combination of two basic probability mass functions, $m_1(x_1)$ and $m_2(x_2)$ in support of a hypothesis A is calculated through a so-called orthogonal addition ($\oplus$) given by:[46]

---

[44] Kari Sentz and Scott Ferson, *Combination of Evidence in Dempster-Shafer Theory*, Sandia National Laboratories, SAND2002-0835, April 2002.

[45] Scott Ferson, Vladik Kreinovich, Lev Ginzburg, Davis S. Myers, and Kari Sentz, *Constructing Probability Boxes and Dempster-Shafer Structures*, Sandia National Laboratories, SAND2002-4015, January 2003.

[46] Sentz and Ferson, 2002.

$$[m_1 \oplus m_2](A) = \frac{\sum_{x_2 \cap x_1 = A} m_1(x_1) \times m_2(x_2)}{1 - \left(\sum_{x_2 \cap x_1 = \emptyset} m_1(x_1) \times m_2(x_2)\right)}$$

where $\emptyset$ is the empty set. The numerator in the above equation is summed over all the elements of the mass functions where the intersection of these functions equate to the hypothesis *A*. The denominator is a normalization factor and is summed over all the elements whose intersection is the empty set.

There are some advantages for using DST, including its ability to assign evidence to more than one hypothesis to account for uncertainty and to provide upper and lower bound probabilities associated with a given claim. Using the same nomenclature as in the above equation, the lower probability limit, also called belief function, *Bel(A)*, and upper probability limit, also called plausibility function, *Pl(A)*, for hypothesis *A* are given by

$$Bel(A) = \sum_{x \sqsubseteq A} m(x)$$

and $Pl(A) = 1 - Bel(\sim A)$, respectively,

where $\sim A$ denotes the negation of the hypothesis A.

The upper and lower probability calculations allow for the development of a so-called probability box (p-box), which essentially involves providing upper and lower probability curves to bound an unknown distribution, and a discretized version can be developed using DST, an example of which is shown in Figure 3.6. Some of the claimed advantages of using p-boxes and DST is that the approach accounts for

- model uncertainty
- non-negligible measurement uncertainty
- small sample size
- inconsistency of input data quality.[47]

Other evidence theory approaches can be used; however, we only considered DST because it appeared to provide the fundamental capability needed for our application, i.e., combining evidence and allowing for uncertainty. In summary, the approach we envision is to quantitively (if possible) estimate the trust level using various layers and then combine them using DST to calculate an overall trust score.

---

[47] Ferson et al., 2003.

Figure 3.6. Dempster-Shafer Theory Probability Box

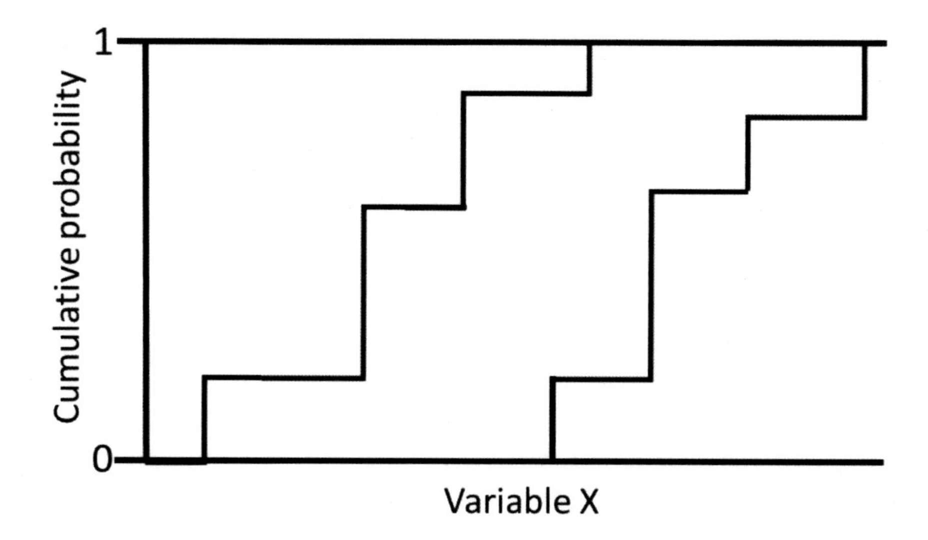

SOURCE: Adapted from Ferson et al., 2003.

The next chapter provides concrete examples of how the resilience assessment framework described in this report would work on two sample USSF mission systems. These illustrative mission analyses omit the modulating effect of trust on the contributions from notional commercial partners.

Chapter 4. Illustrative Mission Analyses

To illustrate how the framework might work in practice, we assessed resilience for two sample USSF missions: DTRN and T-ISR. In this chapter, we examine each mission in turn, laying out assumptions and mission architecture and then testing the resilience of that architecture under varying conditions—focusing primarily on how much resilience changes when commercial services are incorporated into the system architecture.

DTRN Mission Case Study

The DTRN mission is a subcomponent of the USSF GEN effort and is intended to serve as a reliable data communication infrastructure connecting data-acquiring space assets with ground assets that receive, relay, process, and/or store acquired data. The GEN effort includes two other components that interact with DTRN: the Defensive Cyberspace Operations-Space (DCO-S) and the Enterprise Ground Service (EGS). DCO-S is intended to address cybersecurity concerns across the GEN program. EGS provides platforms for computer and data storage for USSF satellite operations.

In this case study, the DTRN plays the connectivity role for all satellite operations across the USSF. We assume there are no inter-satellite links in operation. Connections across ground nodes are assumed to be relayed via internet-based data transfer or any other mode of terrestrial data transfer that EGS can enable. This constraint informs some of our selected system MOPs.

The goals of this assessment were to (1) evaluate the *resilience* of the DTRN mission using the resilience assessment framework; and (2) evaluate how commercial contributions to the DTRN modulate the resilience of the DTRN mission. Another question we explored is the influence of network constraints due to differences in communication protocols for assets in the DTRN. Our goal was to model a method of resilience analysis using the DTRN mission as a test of the resilience assessment framework.

Identifying Mission Architecture and Needs

We first enumerated the relevant Satellite Control Network (SCN) assets and architecture that combine to serve the DTRN mission.[48] The main relevant asset classes, illustrated in Figure 4.1 include satellites (in geostationary orbit [GEO], medium earth orbit [MEO], and low earth orbit [LEO]) and ground antennae. In this example, we show a *simulated* space constellation constructed to reproduce a system summary statistic—the number of network

[48] The SCN was formerly known as the Air Force Satellite Control Network.

contacts per day—reported by the SCN. Future resilience assessments may be made using true and more detailed constellation orbit parameters.

Figure 4.1. DTRN Operational View-1 Depiction

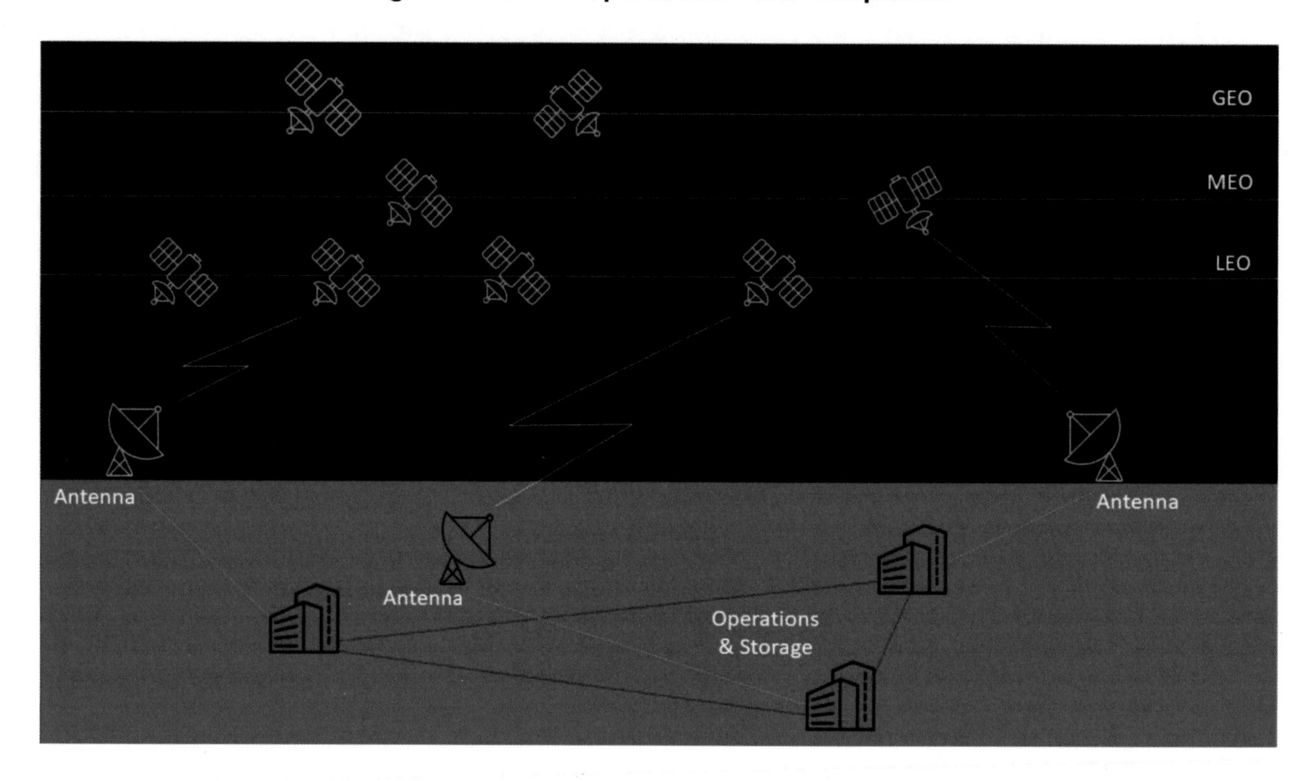

The USSF manages a set of seven ground antenna stations dispersed across the globe as part of its SCN.[49] Commercial entities also operate ground antenna stations whose capacity they can lend to DoD and the USSF under the Commercial Augmentation Service (CAS) program. Our analysis modeled the *USSF-only* architecture using only DoD satellites (a simulated approximation thereof) and the SCN antenna stations. The model of the *USSF + Commercial* architecture included all USSF assets (ground and space), as well as a collection of commercial antenna stations in the CAS program. We also kept track of communication protocol capabilities for all antennae included in the models.

Special Mission Characteristics

We incorporated two other details of the DTRN mission system into our analysis. The first concerns the number of antennae per ground location. The antenna count varied per location for both the SCN and the CAS sites. We can use this variability to determine a nonuniform index of fragility (or sensitivity to degradation) for nodes in the collection of ground sites.

[49] These are named Boss, Cook, Guam, Reef, Hula, Pogo, and Lion.

The second consideration concerns communication bands and protocols. Communication on the DTRN typically happens on the S-, L- and X-bands using the unified S-band (USB) protocol or the Space-to-Ground Link Subsystem (SGLS) protocol. Most CAS operators work with USB, and most government SCN sites use SGLS. Bandwidth efficiency is different between the two protocols.[50] We can use these differences to explore the effects of hypothetical protocol restrictions on DTRN system performance.

Estimating USSF and Commercial Capabilities

Inputting the mission architecture described above, we used AFSIM to simulate the satellite orbits and their ground antenna contacts for a time window of seven days. The AFSIM simulation produced a transmission event time-series database that records all opportunities for data transmission and reception in the architecture for the simulation window. We used this transmission event database to generate MOPs for the baseline USSF-only, the USSF + Commercial, and the degraded mission architectures. We evaluated three main MOPs for the DTRN mission, two of which are illustrated in Figure 4.2:

- *Average connectivity for the antenna collection*: How many satellites are within communication line of sight of an antenna on average? *Average* here refers to a time average over the whole simulation window. At each instant, the mission system has a connectivity distribution,[51] which we summarize (with an ensemble maximum or average) and report. This number is like a system utilization metric. Higher MOP numbers mean that, on average, more satellites are available to each antenna for downlink or uplink, which indicates better system performance.
- *Average time between antenna contact*: What is the average (over satellites) amount of time between antenna connection events? We can think of this measure as a *satellite revisit time* (as opposed to a ground POI revisit period). Lower MOP numbers mean that, on average, a satellite in motion spends less time before reaching an antenna for downlink or uplink, which indicates better system performance.
- *Total contact duration*: What is the total amount of time during which satellites and antennae can communicate? This is an imprecise proxy metric to track an upper bound of the total amount of data that can be transferred on the network.[52] Higher MOP numbers mean that, on average, there is more time available for data transfer, which indicates better system performance.

---

[50] Charles C. Wang, Tien M. Nguyen, and James Yoh, "On the Power Spectral Density of SGLS and USB Waveforms," *1999 IEEE Aerospace Conference Proceedings*, Vol. 2, 1999.

[51] This is the *instantaneous* degree distribution for the dynamic network representation of the mission system. We can summarize this degree distribution with various summary statistics (e.g., maximum, mean, median, minimum).

[52] It is only an imprecise proxy because we do not account for data transfer rates. Data transfer rates would depend on a variety of communication systems and environmental factors.

Figure 4.2. Depiction of Two DTRN Mission Metrics: Satellite Revisit Time and Total Contact Duration

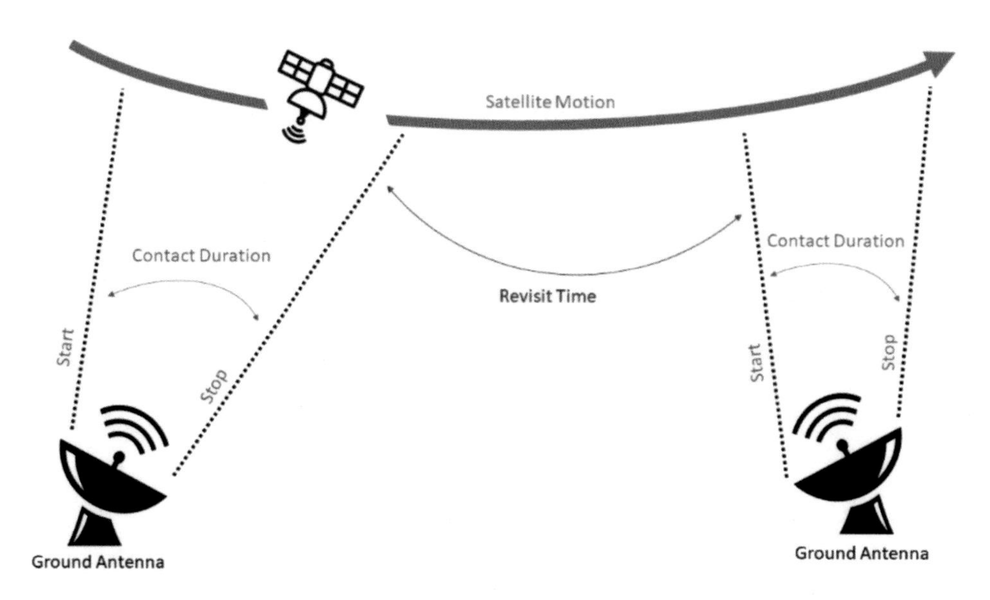

Calculating Resilience Measures

We calculated resilience of the DTRN for all four scenarios shown in Table 2.2: the USSF-only network, undegraded and degraded, and the USSF + Commercial network, undegraded and degraded.

Antenna Connectivity

We first estimated the antenna connectivity MOP comparing the USSF-only and USSF + Commercial mission systems under different levels of degradation of the antenna set. Figure 4.3 summarizes the analytic results using the *average maximum connectivity* metric, which is the time average (over the simulation window) of the instantaneous maximum number of satellites accessible to a single antenna. Degradation was applied to the antenna locations and prioritized by antenna count at each location. Note that the first and leftmost value (0.0) of the graphs in the subsequent figures show the systems' responses for the undegraded scenarios. System responses to higher degradation intensities are shown as lines or numbers extending to the right.

Two findings of this assessment are noteworthy (see Figure 4.3). First, connectivity tends to be higher for the USSF + Commercial mission system than the USSF-only system at all levels of degradation (including the baseline undegraded case). This occurs because there are more ground assets with finer geospatial coverage with the addition of commercial assets and, therefore, more opportunities for satellites to be in line of sight of ground antennae. The second key finding is that system degradation diminishes system performance *at similar rates* for both the USSF-only and the USSF + Commercial systems. In essence, the USSF + Commercial system has a larger performance buffer to resist catastrophic degrading than the USSF-only system.

Figure 4.3. Comparison of Average Antenna Connectivity, by Mission System

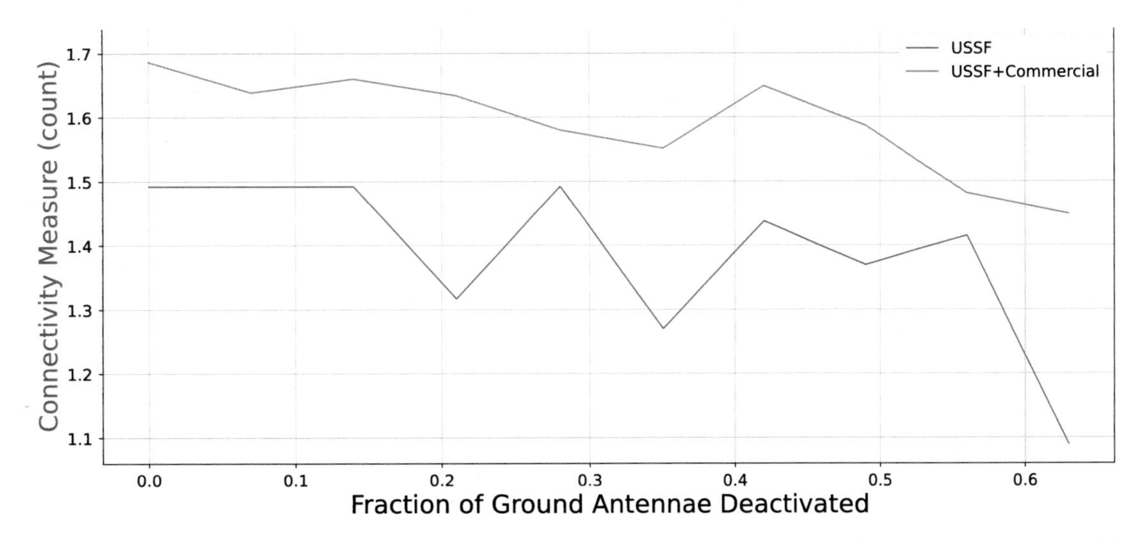

NOTE: The figure shows the results of random asset degradation on the average maximum antenna connectivity MOP on the USSF-only network (identified as USSF in the figure) and the USSF + Commercial network.

Satellite Revisit Time

The two mission configurations also seem to show differential sensitivity to antenna infrastructure degradation for this MOP. Degradation intensity varied from 0 to 85 percent (a dynamic range of 85 percentage points). In Figure 4.4, the USSF + Commercial configuration's mean revisit time increases from about 10 seconds to about 300 seconds for the 85-percent degradation case, while the USSF-only configuration's mean revisit time increases from about 200 seconds to about 1,200 seconds for the same degradation level. This suggests an estimated system elasticity approximately equal to 0.29 ((300–10)/85) for the USSF + Commercial system compared with an estimated system elasticity approximately equal to 0.085 for the USSF-only system on the satellite revisit time MOP. One way of reading this is to say the USSF-only DTRN mission system is over three times more sensitive to changes in the satellite constellation as the USSF + Commercial DTRN mission system.[53] This suggests that the system with the additional commercial capabilities is more resilient. We note that our analysis ran a limited number of instantiations, resulting in a non-monotonous behavior exhibited in some of the figures presenting DTRN mission–related results: For example, in Figure 4.4, the revisit time at 75 percent is higher than that at 85 percent. Figure 4.5 also exhibits this pattern: The degradation at 55 percent is higher than the degradation at 45 percent.

[53] The elasticity factor is only intended as an indicator that we can use for comparison purposes. Another approach would be to use the difference in degraded performance with and without commercial contribution to measure the relative increase in resilience provided by the commercial portion of the architecture, i.e., for the 85-percent degradation case the improvement in performance of the USSF + Commercial architecture versus the USSF-only architecture is more than three times better (1,200 seconds versus 350 seconds).

We evaluated how the *satellite revisit time* (average time between antenna contact) MOP of the USSF-only mission system and the USSF + Commercial mission system fared under antenna infrastructure degradation. Figure 4.4 shows how mean revisit time varies under different intensities of degradation applied to the simulated satellite constellation. We observe results similar to the antenna connectivity comparison: The USSF + Commercial mission system performs better than the USSF-only system at all levels of degradation (including the baseline undegraded case), showing lower revisit times.

Figure 4.4. Comparison of Average Antenna Revisit Time, by Mission System

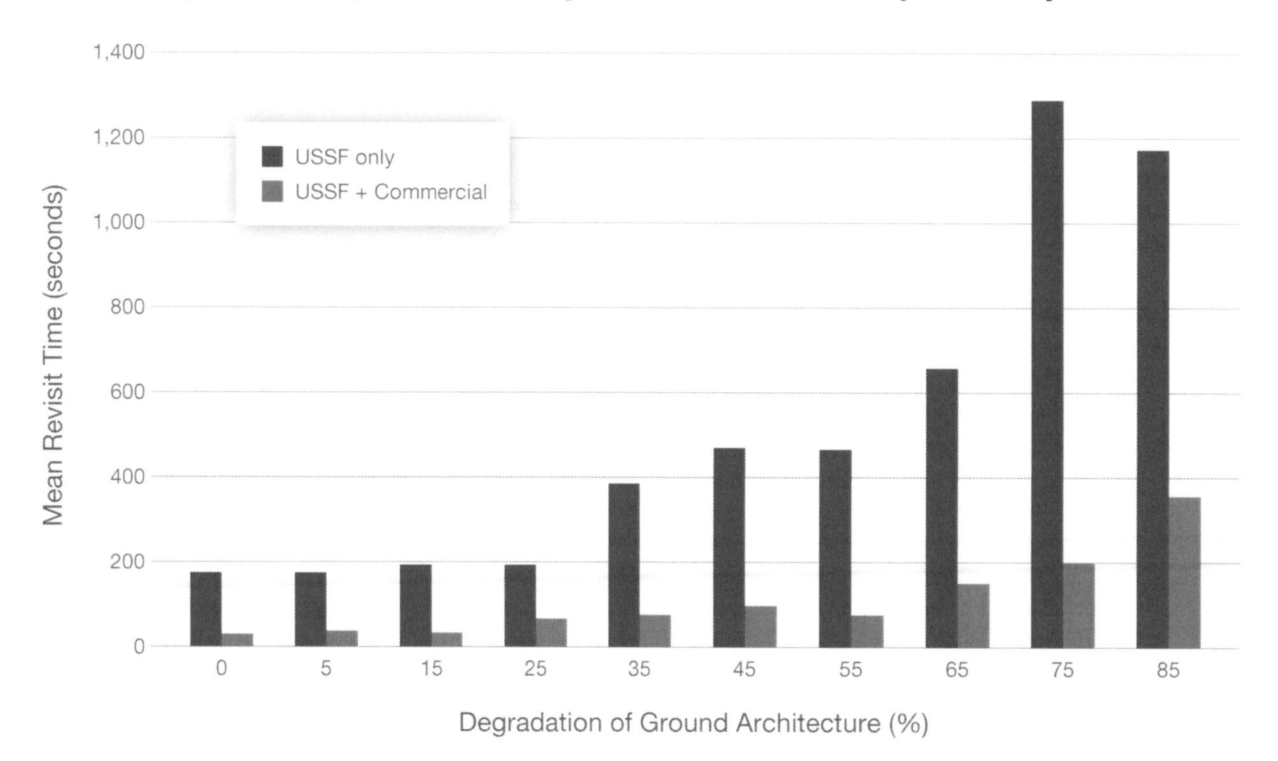

Total Contact Duration

Another view of the resilience of the USSF-only and USSF + Commercial DTRN mission systems is the total contact duration MOP and how it fares under different levels of antenna infrastructure degradation. Figures 4.5 shows these results with a similar conclusion as with the other MOPs: The USSF + Commercial system fares considerably better by providing longer contact duration under all levels of degradation.

Figure 4.5. Comparison of Total Contact Duration, by Mission System

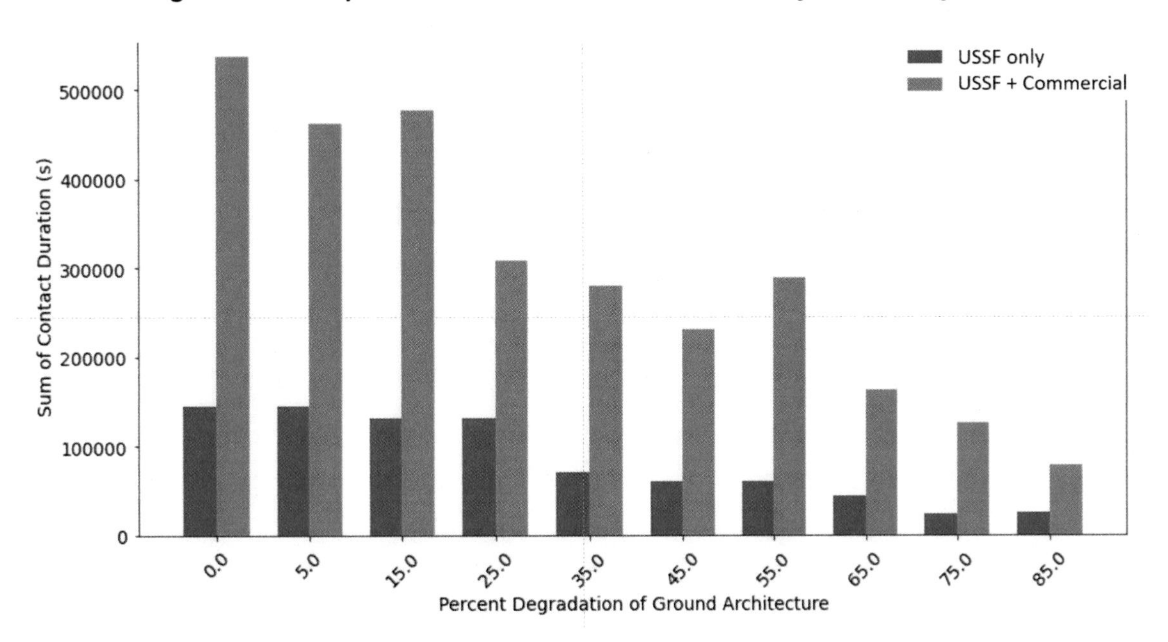

NOTE: s = second.

T-ISR Case Study

We considered a space-based T-ISR mission that performs an occupancy check of 40 air bases using a notional constellation of USSF satellites combined with different constellations of commercial satellites all in LEO. For this application of the framework, we simply describe the mission as the use of imaging satellites to collect information about operationally relevant terrestrial POIs in a timely manner. The time constraints discussed in the case come from the potential need to make quick decisions about new ground missions or to support ongoing ground missions.

Relevant dimensions of this mission include available sensing modalities (such as, EO/IR, radar, hyperspectral), geographic distribution of target locations, the spatiotemporal resolution of information acquisition, and others. Relevant MOPs that we used for this mission include total time needed to acquire all portfolio targets and the revisit rate for coverage of all target locations.

Identifying Mission Architecture and Needs

Our stylized T-ISR mission system consists of two components like the DTRN mission case study: (1) a portfolio of ground POIs to surveil and (2) a collection of space-imaging satellites to do the surveillance. We selected a collection of about 40 geospatially distinct ground targets for the ground POI portfolio. We then used the PlanetScope Dove and SkySat constellation of imaging satellites to simulate the surveillance system. For this stylized experiment, we used the SkySat constellation as a simulacrum, standing in for a notional high-performance USSF-only constellation of imaging satellites. While the Dove constellation simulates a notional portfolio of commercial imaging satellites.

When considering the capabilities of the two Planet Lab constellations under consideration, we referenced the *Planet Imagery Product Specifications* guide and *SkySat Imagery Product Specification* guide.[54] The Dove constellation consists of three generations of these satellites. We assume all satellites are the third-generation SuperDove or PSB.SD. These individual Dove satellites have a ground resolution of approximately 3.7 meters and can only be nadir oriented.

Using the Planet Lab image product size, we estimated the sensor field of view to be approximately 4.13 degrees. Despite the limits to the Dove sensor field of regard, the Dove constellation consists of approximately 120 satellites (as of 2018, estimates were around 130). Compared with the PlanetScope constellation, the SkySat constellation consists of 21 satellites (in 2 different orbits including a polar one) with a ground resolution as high as 0.8 meters. The SkySat sensors also have pointing capabilities, and we estimate their field of regard to be 20 degrees. The constellation is modeled in AFSIM assuming a Walker Delta with the general parameters provided in Table 4.1.

The notional USSF-only constellation modeled in the analysis was assumed to be similar to SkySat's.

Table 4.1. T-ISR Constellation Parameters in AFSIM

Parameter	Dove	SkySat	SkySat (Polar Orbit)
Number of satellites	120	6	15
Satellites per plane	8	3	3
Altitude (km)	475	500	500
Inclination angle (degree)	98	56	98

SOURCE: Features information from Planet, 2020, and Planet, 2018.
NOTE: We used the SkySat constellation as a proxy for the USSF-only constellation, while the Dove constellation represented the commercial contribution.

The goal of this analysis was to simulate how a resilience analysis for a notional T-ISR mission would proceed. The analytic setup allowed us to evaluate the performance and resilience contributions of notional commercial satellite imagery providers to a T-ISR mission for the USSF. We omitted a lot of operational detail in this analysis (e.g., tasking, scheduling, resolution limits, cueing, and extra acquisition modalities). Future specific implementations of the assessment framework can easily include these complexities in the system representation simulations.

---

[54] Planet, *Planet Imagery Product Specifications*, June 2020; Planet, *SkySat Imagery Product Specification*, May 2018.

Estimating USSF and Commercial Capabilities

Inputting the mission architecture specified in the previous section, we used AFSIM to simulate the Dove and SkySat satellite orbits and their ability to observe and image the various ground POIs for a time window of seven days. The AFSIM simulation produced an observation event time-series database that records all opportunities for acquiring images of the marked ground POIs. We used this event database to generate MOPs for the Dove, the SkySat, and the Dove + SkySat architectures, as well as their degraded versions.

For this highly simplified analysis, we evaluated one MOP, *target acquisition time*. This time series shows the number of targets (maximum of about 40) captured within a given time. The steepness of the curve tells us how long it takes for a given T-ISR mission system to acquire some or all of its targets. A steep curve that saturates to its maximum quickly is better than a slow-rising curve or one that never achieves the maximum value (i.e., some targets were not imaged within the time window).

There are other relevant measures to evaluate the average revisit time to a ground site for each type of constellation. Table 4.2 shows the average, minimum, and maximum times for the Dove constellation, SkySat constellation, and the combined Dove + SkySat constellation. The combined Dove + SkySat constellation produces lower average revisit times than either of the individual constellations. The maximum revisit time of the Dove + SkySat constellation also remains lower than that of either the Dove or SkySat constellation.

Table 4.2. Revisit Times for T-ISR Constellation Architectures

	Revisit Time (second)		
Constellation Type	**Average**	**Minimum**	**Maximum**
Dove	370	4	3,266
SkySat	748	4	8,648
Dove + SkySat	248	4	2,090

Calculating Resilience Measures

Figure 4.6 shows the target acquisition performance in the baseline (undegraded) scenarios for the three T-ISR mission systems: a Dove-only system, a SkySat-only system, and a Dove + SkySat system that combines image-acquisition contributions from both constellations. We see that, separately, the Dove and SkySat constellations take about 60,000 to 70,000 seconds to acquire all targets. The combined constellation takes just under 40,000 seconds to complete its acquisition task. Figure 4.7 shows the impact of satellite constellation degradation[55] on the target acquisition performance of the Dove + SkySat mission system. As expected, degradation

[55] Ground node degradation is not available for this mission because the portfolio of ground nodes or POIs are specified by mission requirements. So, they need to be fixed during resilience assessments.

diminishes performance. Further analysis on a more detailed mission may use a similar workflow to evaluate the impact various mission system interventions or augmentations.

Figure 4.6. Comparison of Target Acquisition Times, by Mission System

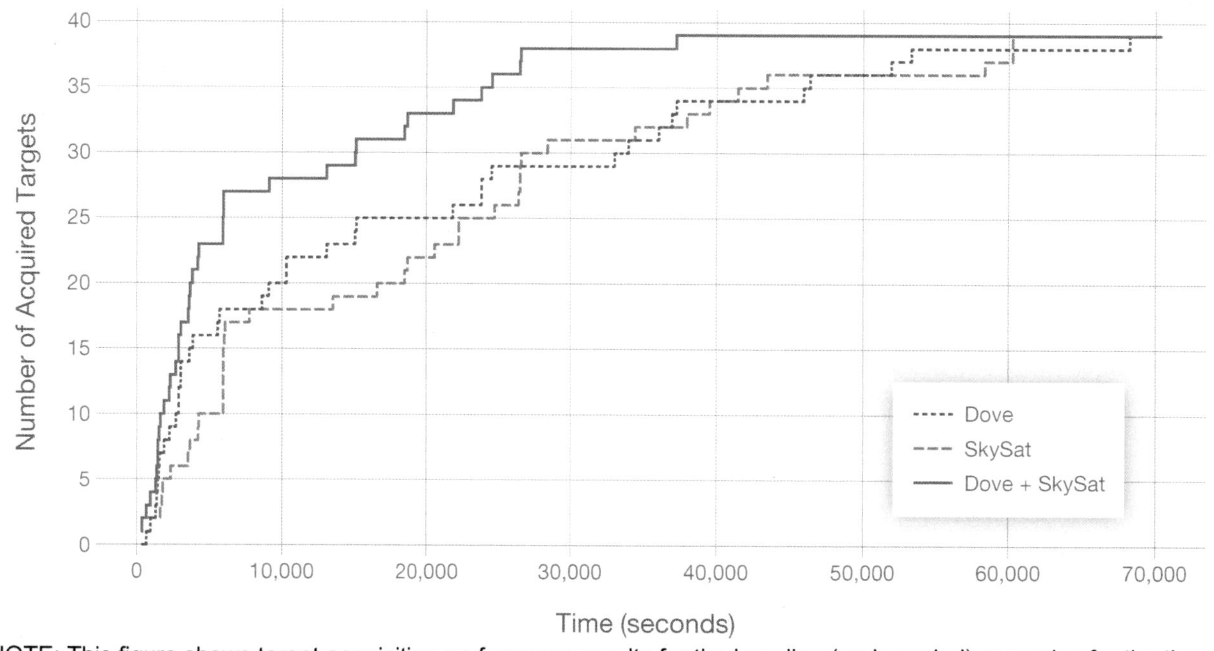

NOTE: This figure shows target acquisition performance results for the baseline (undegraded) scenarios for the three T-ISR mission systems. We used the SkySat constellation as a proxy for the USSF-only constellation, while the Dove constellation represented the commercial contribution.

Figure 4.7. Target Acquisition Performance of Dove + SkySat Constellation, by Degradation Level

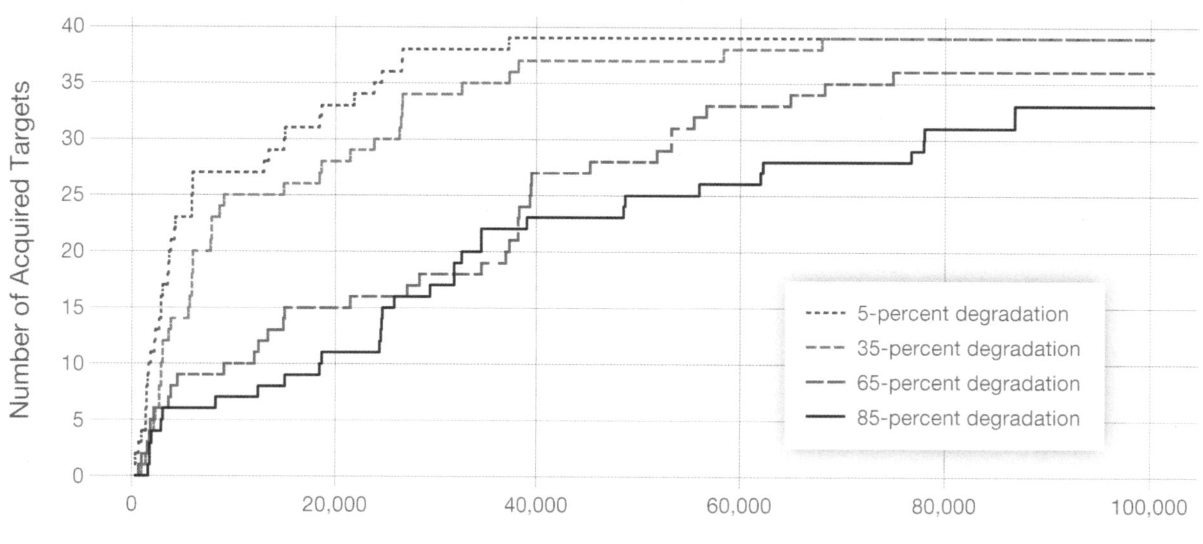

NOTE: This figure shows target acquisition performance of the USSF + Commercial constellation at the indicated degradation levels.

Conclusion

The preceding case analyses demonstrate the value of the resilience assessment framework in the context of two disparate USSF missions. These mission analyses can be extended much further by simulating more interesting or more representative mission systems, identifying other relevant MOPs, or drilling down to calculate detailed estimates of effect sizes. The goal here was to illustrate details about how the framework can be used with simple mission goals, but the capability can be applied to far more complex analyses.

Chapter 5. Summary and Future Directions

The resilience assessment framework described in this report was designed to help the USSF assess the effects of commercial augmentation on the resilience of their mission systems. The framework consists of the following steps: (1) system representation, (2) system evaluation, (3) trust calibration of system subcomponents, and (4) degradation and resilience analyses. The assessment framework was designed to be flexible, modular, and sufficiently representative. The modularity of the framework is especially important because analysts will often want to explore the effects of adding or removing subcomponents of the mission system under analysis.

To help address the complex question of trust calibration, we developed a subframework that enables analysts to combine multiple signals of trustworthiness into an overall trust score. The trust assessment subframework goes beyond other frameworks based on just organization-level assessments of cyber risk, such as CMMC. A variety of fingerprints on sensor data and on the data custody chain can directly inform the trust score derived in this subframework. The technical evaluations discussed in Chapter 3 illustrate the value of ML models for validating sensor fingerprints.

To demonstrate the resilience assessment framework, we ran a streamlined version of the framework on two USSF case-study missions: a DTRN mission and a notional T-ISR mission. These analyses did not include the complexity of the trust calibration step. The results of the analyses proved insightful on the value of commercial augmentation for different dimensions of USSF mission performance. The case studies also helped highlight relevant advantages and limitations, as well as future directions, for expanding the utility of the methodology, which we describe in the remainder of this chapter.

Findings

The assessment framework has several advantages. The main benefits of the approach include the use of a post-processing tool to degrade the expected nominal performance for assessing resilience. The degradation is performed without additional physics-based modeling runs, allowing analysts to consider a variety of degradation cases with minimal effort. We summarize some of the benefits of this framework as follows:

- significant reduction in needed number of physics-based modeling runs
    - post-processing analyses for degradation based on a single physics-based run for each architecture
    - a large number of architectural degradations are quickly assessed
    - multiple types of degradation can be applied, i.e., time based, satellite based, location based

- modularity, i.e., new system subcomponents (real or hypothetical) can often be added to the analysis without having to rerun the initial physics-based model
- resilience measures are directly linked to key operational measures, i.e., MOPs
- generated information supports joint warfighter analysis.

We also identified several key limitations with the framework as follows:

- *Assumption of static system responses*: The degradation analysis currently assumes a *static* response from the mission system when confronted with degradation, i.e., the architecture does not attempt to respond by altering its structure. We are essentially assuming that no immediate, short-term adaptive remedies are triggered when we impose a simulated stress on the mission system. Although this may be representative of current conditions because of limited integration between military and commercial space systems, we assume that such integration will improve, especially as an overall concept of operations (CONOPS) accounting for the integration is developed. In the future, we expect mission assets and mission operators to have some rapid remedies available to address new shocks. One way to interpret the static response assumption is to consider that it aims to identify worst-case outcomes when no remediating actions occur. It is thus a reasonable typical assumption for this type of analysis
- *Customized contributions may require customized representation*: Analysis combining contributions from different architectures sometimes requires customization. For example, analysts may want to automatically consider other ways by which a commercial architecture can contribute to the government one, e.g., tipping/cueing or multi-phenomenology.
- *Complexities in representing specific sensing and transmission physics*: Representing mission-specific sensor constraints (either physical or operational) imposes an added level of complexity. We explored approaches for representing sensor constraints (e.g., on scheduling and sensing modality) with some success. There is further work to be done to characterize the limits of the kinds of sensing physics that can be reliably represented in this kind of dynamic network system representation. There is also further work to be done on how analysts may usefully represent the effects of mission-relevant transient and/or localized phenomena such as space and earth weather.
- *Currently limited to information transmission mission systems*: The current implementation of the framework is only applicable to missions that can be approximated by space and terrestrial nodes that exchange signals or information. In the T-ISR case, space nodes receive electromagnetic signals about ground nodes. In the DTRN case, space and ground nodes swap radio signals. USSF missions that involve the *transmission of matter* may require more elaborate physics to model sufficiently.

- *Selecting and evaluating mission-relevant resilience indices*: The USSF runs and contributes to a diverse set of missions. The relevant performance and resilience metrics will vary with the missions. The validity of a resilience assessment depends crucially on matching the right metrics for the mission. This requires detail and clarity on the specific operational context. The right resilience measure (e.g., as derived directly from mission requirements development) may require more complex modeling than we have shown in the analytic case studies. Future work may broaden the set of tools and methods available for evaluating more complex resilience indices that might be required by other missions under study.

The benefits of the resilience assessment framework greatly outweigh its limitations because, as we demonstrated in Chapter 4, this framework is robust enough to derive mission-relevant insights despite its limitations. Some of the limitations can serve as useful pointers for development opportunities.

Recommendations

We present the following recommendations to extend and develop the utility of the resilience assessment framework in future work.

Extend the Framework to Simulate Responses to Adversarial Targeted Degradation

The most direct extension to the resilience assessment framework is to simulate and evaluate a mission system's response to more targeted adversarial system degradation. The aim would be to approximate the intensity and kinds of effects that a highly motivated and highly informed adversary could have on a mission system. Adversarial degradation contrasts with purely random degradation (a simulation of nonadversarial operational wear and tear) that our work so far applies. Such adversary models can be simulated by an exhaustive Monte Carlo search for the worst-case system degradation, a more theory-driven topological analysis of the mission system's network structure, or some combination of the two approaches. Exhaustive searches over the space of (dynamic) network structures are NP-Hard and likely infeasible for all but the simpler mission systems.[56] Efficient methods for approximating worst-case attacks on mission systems are important topics of further research.

---

[56] NP-Hard is a complexity class of decision problems that the National Institute of Standards and Technology (NIST) defines as "intrinsically harder than those that can be solved by a nondeterministic Turing machine in polynomial time" (Paul E. Black, ed., "NP-hard," in *Dictionary of Algorithms and Data Structures* [online], National Institute of Standards and Technology, January 5, 2021).

Further Develop and Implement the Trust Assessment Subframework

Our work to date recognizes the importance of adopting a calibrated trust approach to information supply chains in a combined USSF + Commercial mission system. We identified several approaches for fingerprinting information artifacts from USSF mission assets for chain-of-custody tracking. And we identified a promising approach for fusing together trust signals (both at the level of the single information artifact and the organization source) into a summary metric of trust. There is further work to do implementing the approach from end to end for a specific mission. There is also further work to do to stress test effective approaches for fusing trust evaluations into the assessment workflow.

Use Improved Representations of USSF Mission Systems in Resilience Analyses

The foundation of the assessment framework is a tractable, flexible, and approximate representation of the mission system under study. We make simplifying approximations of both the mission system's structure and its variation over time (by down-sampling). To achieve more detailed systems evaluation may require a finer approximation of the mission system—such as by representing more complex sensing physics, asset scheduling, cyber threat models, or environmental patterns. There are likely trade-offs between the fineness of the system approximation and the level of detailed resilience analysis possible under this framework. But there are also likely diminishing marginal returns to increasing the quality of system representation for analysis under this framework. Further work needs to be done to better characterize this trade-off space.

Operationalize an Appropriately Tailored Resilience Assessment Framework for USSF Missions

The assessment framework described in this report is designed to be modular and flexible, especially because the motivating question that drove its development was about commercial augmentations to USSF mission systems. There is further work to do to plan out how the USSF may apply this framework, including answering questions about what kinds of systems information the USSF would need to gather from candidate commercial providers to enable the USSF to perform its own mission resilience assessments in house. Commercial provider candidates may need to deliver time-series samples of their observation capability to the USSF and trust-relevant fingerprint data.

Explore Applications of the Framework to Other Missions

The motivating idea behind our resilience assessment framework is not limited in relevance to just missions in space. The framework can apply to any mission that can admit a dynamic network representation as a decent-if-imperfect approximation. As the USSF, and DoD more broadly, continue to engage with the private sector for operational mission support, mission-agnostic resilience assessment frameworks like ours can serve as valuable tools. And, if the USSF and DoD are to rely on third-party information flows to support missions, further development on hybrid calibrated trust evaluation frameworks will be necessary.

Assess What Modifications to the Overall USSF CONOPS Are Required to Effectively Leverage Commercial Capabilities

This should involve assessing the impact of integrating commercial capabilities with USSF mission systems on doctrine, organization, training, materiel, leadership, personnel, facilities, and policy.

Abbreviations

AFSIM	Advanced Framework for Simulation, Integration and Modeling
AI	artificial intelligence
CAS	Commercial Augmentation Service
CMMC	Cybersecurity Maturity Model Certification
CNN	convolutional neural network
CONOPS	concept of operations
DAF	Department of the Air Force
DoD	U.S. Department of Defense
DST	Dempster-Shafer theory
DTRN	data transmit and receive network
EO	electro-optical
FC	fully connected
GEN	Ground Enterprise Next
GEO	geostationary orbit
IR	infrared
JP	Joint Publication
KPP	key performance parameter
LEO	low earth orbit
MEO	medium earth orbit
ML	machine learning
MOP	measure of performance
NASA	National Aeronautics and Space Administration
PAF	Project AIR FORCE
POI	point of interest
PRNU	photon response nonuniformity
SCN	Satellite Control Network
T-ISR	tactical intelligence, surveillance, and reconnaissance
USB	unified S-band
USSF	U.S. Space Force

References

Alaska Satellite Facility, "ALOS AVNIR-2 Ortho Rectified Image Product," dataset, undated. As of January 27, 2023: https://asf.alaska.edu/datasets/optical-data-sets/alos-avnir-2-ortho-rectified-image-product/

Athanasiadou, Eleni, Zeno Geradts, and Erwin Van Eijk, "Camera Recognition with Deep Learning," *Forensic Sciences Research*, Vol. 3, No. 3, July 2018.

Black, Paul E., ed., "NP-hard," in *Dictionary of Algorithms and Data Structures* [online], National Institute of Standards and Technology, January 5, 2021. As of January 11, 2023: https://xlinux.nist.gov/dads/HTML/nphard.html

Burch, Ron, *Resilient Space Systems Design: An Introduction*, CRC Press, 2019.

Carley, Kathleen M., Jana Diesner, Jeffrey Reminga, and Maksim Tsvetovat, "Toward an Interoperable Dynamic Network Analysis Toolkit," *Decision Support Systems*, Vol. 43, No. 4, August 2007.

Clive, Peter D., Jeffrey A. Johnson, Michael J. Moss, James M. Zeh, Brian M. Birkmire, and Douglas D. Hodson, "Advanced Framework for Simulation, Integration and Modeling (AFSIM)," in *Proceedings of the International Conference on Scientific Computing (CSC)*, Steering Committee of the World Congress in Computer Science, Computer Engineering and Applied Computing (WorldComp), 2015.

Cozzolino, Davide, Francesco Marra, Diego Gragnaniello, Giovanni Poggi, and Luisa Verdoliva, "Combining PRNU and Noiseprint for Robust and Efficient Device Source Identification," *EURASIP Journal on Information Security*, 2020.

Cozzolino, Davide, and Luisa Verdoliva, "Noiseprint: A CNN-Based Camera Model Fingerprint," *IEEE Transaction on Information Forensics and Security*, Vol. 15, No. 1, 2020.

Defense Innovation Unit, "xView Challenge Series," webpage, undated. As of January 18, 2023: https://www.diu.mil/ai-xview-challenge#xview2

Dempster, Arthur P., "A Generalization of Bayesian Inference," *Journal of the Royal Statistical Society: Series B (Methodological)*, Vol. 30, No. 2, 1968.

Dirik, Ahmet Emir, and Ahmet Karakucuk, *Anonymization System and Method for Digital Images*, patent application, World Intellectual Property Organization International Bureau, International Publication Number WO2014/163597A2, October 9, 2014.

Dreyer, Paul, Krista S. Langeland, David Manheim, Gary McLeod, and George Nacouzi, *RAPAPORT (Resilience Assessment Process and Portfolio Option Reporting Tool): Background and Method*, RAND Corporation, RR-1169-AF, 2016. As of November 16, 2020:
https://www.rand.org/pubs/research_reports/RR1169.html

Earthdata, "Data Processing Levels," National Aeronautics and Space Administration, last updated July 13, 2021. As of August 29, 2021:
https://earthdata.nasa.gov/collaborate/open-data-services-and-software/data-information-policy/data-levels

Ferson, Scott, Vladik Kreinovich, Lev Ginzburg, Davis S. Myers, and Kari Sentz, *Constructing Probability Boxes and Dempster-Shafer Structures*, Sandia National Laboratories, SAND2002-4015, January 2003.

Fridrich, Jessica, Miroslav Goljan, and Jan Lukáš, *Method and Apparatus for Identifying an Imaging Device*, U.S. Patent 7,787,030 B2, August 31, 2010.

Geraldo, Steven, and Randy Woods, "CISA's Cybersecurity Evaluation Tool (CSET)," presentation at the Chemical Security Seminars Virtual Expo, December 16, 2020.

Joint Publication 3-14, *Space Operations*, Joint Chiefs of Staff, October 26, 2020.

Kim, Yool, George Nacouzi, Mary Lee, Brian Dolan, Krista Romita Grocholski, Emmi Yonekura, Moon Kim, Thomas Light, and Raza Khan, *Leveraging Commercial Space Capabilities to Enhance the Space Architecture of the U.S. Department of Defense*, RAND Corporation, 2022, Not available to the general public.

Kinser, Sean, Pete de Graaf, Matthew Stein, Frank Hughey, Rob Roller, David Voss, and Amanda Salmoiraghi, "Scoring Trust Across Hybrid-Space: A Quantitative Framework Designed to Calculate Cybersecurity Ratings, Measures, and Metrics to Inform a Trust Score," Small Satellite Conference, 2020.

Lukáš, Jan, Jessica Fridrich, and Miroslav Goljan, "Digital Camera Identification from Sensor Pattern Noise," *IEEE Transactions on Information Forensics and Security*, Vol. 1, No. 2, June 2006.

Nacouzi, George, Osonde A. Osoba, Jeff Hagen, Jonathan Tran, Christopher Lynch, Mel Eisman, Li Ang Zhang, Charlie Barton, Marissa Herron, and Yool Kim, *Commercial Space Services: An Opportunity for U.S. Space Force to Increase its Mission Resilience*, RAND Corporation, forthcoming, Not available to the general public.

Nevell, David A., Simon R. Maskell, Paul R. Horridge, and Hayleigh L. Barnett, "Fusion of Data from Sources with Different Levels of Trust," *2010 13th International Conference on Information Fusion*, 2010.

Office of the Assistant Secretary of Defense for Homeland Defense and Global Security, *Space Domain Mission Assurance: A Resilience Taxonomy*, September 2015.

Planet, *SkySat Imagery Product Specification*, May 2018.

Planet, *Planet Imagery Product Specifications*, June 2020.

Sentz, Kari, and Scott Ferson, *Combination of Evidence in Dempster-Shafer Theory*, Sandia National Laboratories, SAND2002-0835, April 2002.

Shafer, Glenn, *A Mathematical Theory of Evidence*, Princeton University Press, 1976.

Shaw, Graeme Barrington, *The Generalized Information Network Analysis Methodology for Distributed Satellite Systems*, dissertation, Massachusetts Institute of Technology, 1999.

Shaw, Graeme B., David W. Miller, and Daniel E. Hastings, "Development of the Quantitative Generalized Information Network Analysis Methodology for Satellite Systems," *Journal of Spacecraft and Rockets,* Vol. 38, No. 2, March–April 2001.

Wang, Charles C., Tien M. Nguyen, and James Yoh, "On the Power Spectral Density of SGLS and USB Waveforms," *1999 IEEE Aerospace Conference Proceedings*, Vol. 2, 1999.

Welburn, Jonathan William, Aaron Strong, Florentine Eloundou Nekoul, Justin Grana, Krystyna Marcinek, Osonde A. Osoba, Nirabh Koirala, and Claude Messan Setodji, *Systemic Risk in the Broad Economy: Interfirm Networks and Shocks in the U.S. Economy*, RAND Corporation, RR-4185-RC, 2020. As of August 29, 2021: https://www.rand.org/pubs/research_reports/RR4185.html

Zhao, Bo, Shaozeng Zhang, Chunxue Xu, Yifan Sun, and Chengbin Deng, "Deep Fake Geography? When Geospatial Data Encounter Artificial Intelligence," *Cartography and Geographic Information Science*, Vol. 48, No. 4, July 2021.